OLED Display Fundamentals and Applications

Wiley-SID Series in Display Technology

Series Editors:

Anthony C. Lowe, The Lambent Consultancy, Braishfield, UK
Ian Sage, Abelian Services, Malvern, UK

A complete list of the titles in this series appears at the end of this volume.

OLED Display Fundamentals and Applications

Second Edition

Takatoshi Tsujimura

Registered Office
John Wiley & Sons, Inc., 111 River Street, Hoboken, NJ 07030, USA

Editorial Office
111 River Street, Hoboken, NJ 07030, USA

For details of our global editorial offices, customer services, and more information about Wiley products visit us at www.wiley.com.

Wiley also publishes its books in a variety of electronic formats and by print-on-demand. Some content that appears in standard print versions of this book may not be available in other formats.

Library of Congress Cataloguing-in-Publication Data

Names: Tsujimura, Takatoshi.
Title: OLED display fundamentals and applications / Takatoshi Tsujimura.
Description: Second edition. | Hoboken, New Jersey : John Wiley & Sons Inc.,
  2017. | Series: Wiley series in display technology | Includes
  bibliographical references and index.
Identifiers: LCCN 2016042831| ISBN 9781119187318 (hardback) | ISBN
  9781119187486 (epub) | ISBN 9781119187325 (epdf)
Subjects: LCSH: Flat panel displays. | Electrtoluminescent devices. | Organic
  semiconductors. | Light emitting diodes. | BISAC: TECHNOLOGY & ENGINEERING
  / Electronics / General.
Classification: LCC TK7882.I6 T84 2017 | DDC 621.3815/422–dc23 LC record available at
  https://lccn.loc.gov/2016042831

Cover image: © Blend Images - Colin Anderson/ Gettyimages
Cover design by Wiley

Set in 10/12pt Warnock by SPi Global, Chennai, India

V10006199_111918

Contents

About the Author   *xi*

Preface   *xiii*

Series Editor's Foreword to the Second Edition   *xv*

1   Introduction   *1*
         References   *5*

2   OLED Devices   *7*
2.1   OLED Definition   *7*
2.1.1   History of OLED Research and Development   *7*
2.1.2   Luminescent Effects in Nature   *8*
2.1.3   Difference Between OLED, LED, and Inorganic ELs   *11*
2.1.3.1 Inorganic EL   *11*
2.1.3.2 LED   *11*
2.2   Basic Device Structure   *12*
2.3   Basic Light Emission Mechanism   *14*
2.3.1   Potential Energy of Molecules   *14*
2.3.2   Highest Occupied and Lowest Unoccupied Molecular Orbitals (HOMO and LUMO)   *15*
2.3.3   Configuration of Two Electrons   *17*
2.3.4   Spin Function   *20*
2.3.5   Singlet and Triplet Excitons   *20*
2.3.6   Charge Injection from Electrodes   *24*
2.3.6.1 Charge Injection by Schottky Thermionic Emission   *25*
2.3.6.2 Tunneling Injection   *28*
2.3.6.3 Vacuum-Level Shift   *28*
2.3.7   Charge Transfer and Recombination   *29*
2.3.7.1 Charge Transfer Behavior   *29*
2.3.7.2 Space-Charge-Limited Current   *29*
2.3.7.3 Poole–Frenkel conduction   *32*
2.3.7.4 Recombination and Generation of Excitons   *33*

2.4	Emission Efficiency	36
2.4.1	Internal/External Quantum Efficiency	36
2.4.2	Energy Conversion and Quenching	37
2.4.2.1	Internal Conversion	37
2.4.2.2	Intersystem Crossing	37
2.4.2.3	Doping	38
2.4.2.4	Quenching	40
2.4.3	Outcoupling Efficiency of OLED Display	42
2.4.3.1	Light Output Distribution	42
2.4.3.2	Snell's Law and Critical Angle	43
2.4.3.3	Loss Due to Light Extraction	44
2.4.3.4	Performance Enhancement by Molecular Alignment	45
2.5	Lifetime and Image Burning	46
2.5.1	Lifetime Definitions	46
2.5.2	Degradation Analysis and Design Optimization	47
2.5.3	Degradation Measurement and Mechanisms	50
2.5.3.1	Acceleration Factor and Temperature Contribution	50
2.5.3.2	Degradation Mechanism Variation	50
2.6	Technologies to Enhance the Device Performance	51
2.6.1	Thermally Activated Delayed Fluorescence	51
2.6.2	Other Types of Excited States	53
2.6.2.1	Excimer and Exciplex	53
2.6.2.2	Charge-Transfer Complex	53
2.6.3	Charge Generation Layer	54
	References	56
3	OLED Manufacturing Process	61
3.1	Material Preparation	61
3.1.1	Basic Material Properties	61
3.1.1.1	Hole Injection Material	61
3.1.1.2	Hole Transportation Material	62
3.1.1.3	Emission Layer Material	62
3.1.1.4	Electron Transportation Material and Charge Blocking Material	63
3.1.2	Purification Process	67
3.2	Evaporation Process	68
3.2.1	Principle	68
3.2.2	Evaporation Sources	72
3.2.2.1	Resistive Heating Method	72
3.2.2.2	Electron Beam Evaporation	75
3.2.2.3	Monitoring Thickness Using a Quartz Oscillator	76
3.3	Encapsulation	79
3.3.1	Dark Spot and Edge Growth Defects	79

3.3.2   Light Emission from the Bottom and Top of the OLED Device   *80*
3.3.3   Bottom Emission and perimeter sealing   *81*
3.3.4   Top Emission   *82*
3.3.5   Encapsulation Technologies and Measurement   *83*
3.3.5.1   Thin-Film Encapsulation   *84*
3.3.5.2   Face Sealing Encapsulation   *87*
3.3.5.3   Frit Encapsulation   *88*
3.3.5.4   WVTR Measurement   *88*
3.4   Problem Analysis   *91*
3.4.1   Ionization Potential Measurement   *91*
3.4.2   Electron Affinity Measurement   *92*
3.4.3   HPLC Analysis   *93*
3.4.4   Cyclic Voltammetry   *94*
           References   *96*

4   OLED Display Module   *99*
4.1   Comparison Between OLED and LCD Modules   *99*
4.2   Basic Display Design and Related Characteristics   *101*
4.2.1   Luminous Intensity, Luminance, and Illuminance   *101*
4.2.1.1   Luminous Intensity   *101*
4.2.1.2   Luminance   *102*
4.2.1.3   Illuminance   *103*
4.2.1.4   Metrics Summary   *104*
4.2.1.5   Helmholtz–Kohlrausch Effect   *106*
4.2.2   OLED Current Efficiencies and Power Efficacies   *106*
4.2.3   Color Reproduction   *109*
4.2.4   Uniform Color Space   *115*
4.2.5   White Point Determination   *116*
4.2.6   Color Boost   *119*
4.2.7   Viewing Condition   *120*
4.3   Passive-Matrix OLED Display   *121*
4.3.1   Structure   *121*
4.3.2   Pixel Driving   *122*
4.4   Active-Matrix OLED Display   *125*
4.4.1   OLED Module Components   *125*
4.4.2   Two-Transistor One-Capacitor (2T1C) Driving Circuit   *127*
4.4.3   Ambient Performance   *136*
4.4.3.1   Living Room Contrast Ratio   *136*
4.4.3.2   Chroma Reduction Due to Ambient Light   *137*
4.4.4   Subpixel Rendering   *138*
           References   *139*

5	OLED Color Patterning Technologies	*143*
5.1	Color-Patterning Technologies	*143*
5.1.1	Shadow Mask Patterning	*143*
5.1.1.1	Shadow Mask Process	*143*
5.1.1.2	Blue Common Layer	*146*
5.1.1.3	Polychromatic Pixel	*147*
5.1.2	White + Color Filter Patterning	*148*
5.1.3	Color Conversion Medium (CCM) Patterning	*149*
5.1.4	Laser-Induced Thermal Imaging (LITI) Method	*149*
5.1.5	Radiation-Induced Sublimation Transfer (RIST) Method	*151*
5.1.6	Dual-Plate OLED Display (DOD) Method	*152*
5.1.7	Other Methods	*153*
5.2	Solution-Processed Materials and Technologies	*153*
5.3	Next-Generation OLED Manufacturing Tools	*158*
5.3.1	Vapor Injection Source Technology (VIST) Deposition	*158*
5.3.2	Hot-Wall Method	*163*
5.3.3	Organic Vapor-Phase Deposition (OVPD) Method	*164*
	References	*165*
6	TFT and Driving for Active-Matrix Display	*167*
6.1	TFT Structure	*167*
6.2	TFT Process	*169*
6.2.1	Low-Temperature Polysilicon Process Overview	*169*
6.2.2	Thin-Film Formation	*172*
6.2.3	Patterning Technique	*173*
6.2.4	Excimer Laser Crystallization	*177*
6.3	MOSFET Basics	*180*
6.4	LTPS-TFT-Driven OLED Display Design	*183*
6.4.1	OFF Current	*183*
6.4.2	Driver TFT Size Restriction	*184*
6.4.3	Restriction Due to Voltage Drop	*185*
6.4.4	LTPS-TFT Pixel Compensation Circuit	*190*
6.4.4.1	Voltage Programming	*190*
6.4.4.2	Current Programming	*192*
6.4.4.3	External Compensation Method	*193*
6.4.4.4	Digital Driving	*194*
6.4.5	Circuit Integration by LTPS-TFT	*197*
6.5	TFT Technologies for OLED Displays	*200*
6.5.1	Selective Annealing Method	*200*
6.5.1.1	Sequential Lateral Solidification (SLS) Method	*200*
6.5.1.2	Selective Annealing by Microlens Array	*200*
6.5.2	Microcrystalline and Superamorphous Silicon	*202*
6.5.3	Solid-Phase Crystallization	*205*

6.5.3.1   MIC and MILC Methods   *205*
6.5.3.2   AMFC Method   *205*
6.5.4   Oxide Semiconductors   *207*
          References   *210*

7   OLED Television Applications   *215*
7.1   Performance Target   *215*
7.2   Scalability Concept   *217*
7.2.1   Relationship between Defect Density and Production Yield   *217*
7.2.1.1   Purpose of Yield Simulation   *217*
7.2.1.2   Defective Pixel Number Estimation Using the Poisson
          Equation   *217*
7.2.2   Scalable Technology   *217*
7.2.2.1   Scalability   *218*
7.3   Murdoch's Algorithm to Achieve Low Power and Wide Color
       Gamut   *219*
7.3.1   A Method for Achieving Both Low Power and Wide Color
        Gamut   *219*
7.3.2   RGBW Driving Algorithm   *221*
7.4   An Approach to Achieve 100% NTSC Color Gamut With Low Power
       Consumption Using White + Color Filter   *224*
7.4.1   Consideration of Performance Difference between W-RGB and
        W-RGBW Method   *224*
7.4.1.1   Issues of White + Color Filter Method for Large Displays   *224*
7.4.1.2   Analysis of W-RGBW Approach to Circumvent Its Trade-off
          Situation   *224*
7.4.1.3   Design of a Prototype to Demonstrate That Low Power
          Consumption Can Be Achieved with Large Color Gamut   *229*
7.4.1.4   Product-Level Performance Demonstration by the Combination of
          Scalable Technologies   *230*
          References   *233*

8   New OLED Applications   *235*
8.1   Flexible Display/Wearable Displays   *235*
8.1.1   Flexible Display Applications   *235*
8.1.2   Flexible Display Substrates   *235*
8.1.3   Laser Liftoff Process   *236*
8.1.4   Barrier Technology for Flexible Displays   *240*
8.1.5   Organic TFTs for Flexible Displays   *241*
8.1.5.1   Organic Semiconductor Materials   *242*
8.1.5.2   Organic TFT Device Structure and Processing   *243*
8.1.5.3   Organic TFT Characteristics   *245*
8.2   Transparent Displays   *245*

8.3   Tiled Display   247
8.3.1   Passive-Matrix Tiling   247
8.3.2   Active-Matrix Tiling   248
           References   252

9   OLED Lighting   255
9.1   Performance Improvement of OLED Lighting   255
9.2   Color Rendering Index   257
9.3   OLED Lighting Requirement   259
9.3.1   Correlated Color Temperature (CCT)   260
9.3.2   Other Requirements   262
9.4   Light Extraction Enhancement of OLED Lighting   262
9.4.1   Various Light Absorption Mechanisms   262
9.4.2   Microlens Array Structure   266
9.4.3   Diffusion Structure   266
9.4.4   Diffraction Structure   268
9.4.5   Reduction of Plasmon Absorption   268
9.4.5.1   Plasmonic Loss Mechanism   268
9.5   Color Tunable OLED Lighting   269
9.6   OLED Lighting Design   272
9.6.1   Resistance Reduction   272
9.6.2   Current Reduction   272
9.7   Roll-to-Roll OLED Lighting Manufacturing   273
           References   275

           Appendix   277

           Index   281

About the Author

Takatoshi Tsujimura joined IBM Japan for TFT-LCD development and was selected as one of the "10 best engineers/researchers in the 10 best Japanese companies" by Nikkei Electronics Magazine. He demonstrated OLED's capability to be applied to large television by the world's largest 20-in. demonstration and received SID Special Recognition Award in 2008. He moved to Kodak as a director and developed 100% NTSC white + color filter OLED display with less power consumption than LCDs, which has become industry-standard technology for OLED TV over 50 in. He is currently general manager of OLED business unit, Konica Minolta Inc. He received SID Fellow Award in 2013. He is an SID executive and is a former SID Japan chapter chair. He received PhD in materials science and engineering.

Dr. Tsujimura holds 144 worldwide registered patents and 7 publications including the following:

T. Tsujimura, *OLED Display Fundamentals and Application*, SID-Wiley Series in Display Science, ISBN: 978-1-118-14051-2 (2012)

T. Tsujimura, *OLED Overview (Japanese)*, Sangyo Tosho, ISBN: 978–4782855560 (2010)

T. Tsujimura, *OLED display overview (Korean)*, Hantee Media, ISBN: 8964211766 (2013)

T. Tsujimura, *OLED display overview (Chinese)*, Publication House of Electronics Industry, ISBN: 9787121262814 (2015)

T. Tsujimura, *Technology and Material Development*, CMC Technical Library (2010)

T. Tsujimura, *OLED Material Technology*, CMC Publishing (2004)

T. Tsujimura, *OLED Handbook*, Realize Science and Engineering (2004).

Preface

In the 30 years that have passed since the release of the "first OLED paper," there have been many publications promoting organic light-emitting diode (OLED) displays as a superior technology. Although many aspects of OLED performance are excellent, such as the ultimate high contrast provided by its self-emissive mechanism, it has not been an easy path to achieve wide acceptance in the industry. A decade ago, I was often told by my customers that they totally agree with the beauty of OLED screens, but they are not affordable for common display applications. It is always the case that emerging technology faces difficulty in the wide-scale adoption, in spite of its inherent advantages. The well-established liquid-crystal display (LCD) has been manufactured for a very long time, so it has already been progressively engineered to improve its shortcomings, such as viewing angle, response time, and cost. It was not easy for OLED to overcome such a situation.

What triggered OLED to become so popular was the rapid growth of smart phones. The touchscreen interface paired very well with the OLED screen, as OLED did not show any "touch mura," a smear-like contrast reduction caused by flow-induced liquid crystal deformation. LCD, in a short time, improved the touch mura, but OLED makers utilized this momentary advantage to obtain a chance of growth. From this penetration of OLED display screens, end customers had a chance to recognize the superior display image quality of OLED, especially the beauty of its high contrast. This fortunate circumstance was a major benefit to OLED technology.

More recently, OLED has also penetrated the large television market, taking advantage of the high color gamut white + color filter method we developed, which promises high-yield manufacturing. OLED technology offers many important features for good television, so we expect it will gradually become more popular as the manufacturing cost goes down.

It is still a very important phase for OLED technology to become more widely used. The purpose of this book is to provide necessary information for everyone who is involved with this great technology, so that such knowledge can lead to further improvements. I also hope the contents will help readers to hit upon new great applications, taking advantage of the features of OLED devices.

Takatoshi Tsujimura

Series Editor's Foreword to the Second Edition

In the 4 years since the first edition of Dr Tsujimura's book was published, the status of OLED displays has undergone a transformation. Their development can be exemplified by two products; the introduction of OLED televisions with diagonal screen sizes and resolution fully matching those of LCD sets but with curved screens represented a striking innovation which at that time could not be provided by LCD panels. Many commentators were surprised by the speed with which LCD manufacturers responded to this challenge. On the other hand, OLED panels on flexible substrates provide thin, conformable displays offering excellent compatibility with touchscreen technologies, which many users regard as a benchmark for display performance on mobile electronic devices.

OLED devices are therefore showing their ability to drive customer expectations for display performance. OLED displays provide a performance lead in terms of thin profile, response speed, and black level dominated dark room contrast, which appear difficult to challenge, while brightness, stability, and manufacturing cost are relatively weaker points. Meanwhile, competition between OLED and LCD manufacturers has stimulated innovation in both technologies, driving improvement in key areas such as power efficiency, color gamut, and resolution. As self-emissive devices, OLEDs are also generating widespread interest for lighting applications where their ability to offer a large area source and to achieve high-quality color rendering and tunable color temperature promises an excellent quality of illumination.

This second edition of Dr Tsujimura's book recognizes all these advances through updating and revision of his earlier material, and the addition of extensive new sections covering advances in the technological exploitation of OLEDs. Some of the most prominent of these include approaches to improved power efficiency and color gamut for OLED television, and how to combine both of these key advances; materials and manufacturing methods for flexible OLED displays; roll-to-roll manufacture of OLED lighting panels; and structures of transparent OLED panels. The treatment of a number of the underlying scientific principles involved in device operation and efficiency is also expanded, with new material on luminescence mechanisms, light trapping and extraction, the behavior of various pixel driver circuits, and numerous other topics. It is our hope that these changes will both maintain the currency of the volume and increase its value to a broad range of users.

Dr Tsujimura brings to his subject, the experience and background knowledge gained through his long career researching and developing active matrix technologies, liquid crystal displays, and OLEDs. His knowledge and enthusiasm for his subject are clear in this book, which I believe will continue to provide a most valuable resource both for those working on OLED technologies and their applications, and for scientists and engineers who wish to increase their knowledge of this important field.

Malvern, UK

Ian Sage, Series Editor

1

Introduction

The basic structure of organic light-emitting diodes (OLEDs) was reported by Tang and Van Slyke at Eastman Kodak in 1987 [1]. This was a groundbreaking study and was later referred to as the "first OLED paper." Now, almost 30 years later, there is a large market for OLED devices. The first OLED product was developed by Pioneer for car audio. Then the first mass production of AMOLED by SK display (a joint manufacturing venture by Eastman Kodak and Sanyo Electric) for Kodak's LS633 digital camera (Figure 1.1) accelerated the use of OLED for display applications.

This was followed by the widescale development of many other OLED-based products, including cellular phones (Figure 1.2), smart watches (Figure 1.3), audio players (Figure 1.4), and portable global positioning satellite (GPS) devices, which now provide high-resolution displays in brilliant, multitone colors.

Larger-display products have also been introduced on the market, such as those shown in Figure 1.5. Much larger prototypes have also been developed (Figure 1.6). Because of their superior features such as slim flat-screen design and aesthetically pleasing screen image, and due to high-contrast image signal emission and very good response time, the current state of the art of OLED television technology that has debuted in the marketplace is indeed groundbreaking [2].

The main objective of this book is to explain the basics and application of this promising technology from various perspectives.

OLED Display Fundamentals and Applications, Second Edition. Takatoshi Tsujimura.
© 2017 John Wiley & Sons, Inc. Published 2017 by John Wiley & Sons, Inc.

Figure 1.1 The first active-matrix OLED display product on the market (Kodak LS633 digital camera).

Figure 1.2 Example of a cellular phone using active-matrix OLED (AMOLED) (Galaxy S7 smartphone by Samsung).

Figure 1.3 Example of a smart watch using active-matrix OLED (AMOLED) (Apple Watch Series 2 by Apple).

Figure 1.4 Example of an audio player using active-matrix OLED (AMOLED) (Sony Walkman NW-X-1050).

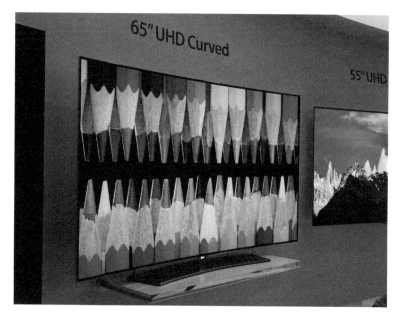

Figure 1.5 Example of a large television using active-matrix OLED (AMOLED) (65-in. curved OLED TV demonstrated in SID2015 by LG display).

Figure 1.6 111-in. dual-sided flexible OLED television prototype (LG Display at IMID2016).

References

1 C. W. Tang and S. A. Van Slyke, Organic electroluminescent diodes, *Appl. Phys. Lett.* **51**(12):913–915 (1987).

2 T. Tsujimura, W. Zhu, S. Mizukoshi, N. Mori, M. Yamaguchi, K. Miwa, S. Ono, Y. Maekawa, K. Kawabe, M. Kohno, and K. Onomura, Advancements and outlook of high performance active-matrix OLED displays, *SID 2007 Digest*, 2007, p. 84.

2

OLED Devices

2.1 OLED DEFINITION

2.1.1 History of OLED Research and Development

Before any in-depth discussion of OLED display structure, let us consider the initial origins of OLED technology, which are based on early observations of electroluminescence (EL). In the early 1950s, a group of investigators at Nancy University in France applied high-voltage alternating-polarity fields in air to thin films of cellulose or cellophane containing deposited or dissolved acridine orange and quinacrine, and observed light emission [1]. One mechanism identified in these reaction processes involved excitation of electrons. Then in 1960, a team of investigators at New York University (NYU) made ohmic (a nonrectifying charge injection, which shows linear current–voltage relationship) dark-injecting electrode contacts to organic crystals and described the necessary workfunctions (energy requirements) for hole and electron-injecting electrode contacts [2]. These contacts are the source of charge injection in all present-day OLED devices. The same NYU group also studied direct-current (DC) EL in vacuo on a single pure anthracene crystal and tetracene-doped anthracene crystals in the presence of a small-area silver electrode at 400 V [3]. The proposed mechanism for this reaction was termed *field-accelerated electron excitation of molecular fluorescence.* The NYU group later observed that in the absence of an external electric field, the EL in anthracene crystals results from recombination of electron and hole and that the conducting-level energy of anthracene is higher than the exciton energy level [4].

Because of the association between EL and later OLED development on the basis of these and other early EL studies, the term *organic EL* gradually emerged and is still used today. EL includes two basic phenomena:

1. Light emission due to the presence of excited molecules caused by accelerated electrons (i.e., electrons that are accelerated to higher energy levels)
2. Light emission due to electron–hole recombination, as in all light-emitting diodes (LEDs).

OLED Display Fundamentals and Applications, Second Edition. Takatoshi Tsujimura.
© 2017 John Wiley & Sons, Inc. Published 2017 by John Wiley & Sons, Inc.

Table 2.1 Differences between Liquid Crystal and OLED Displays

Parameter	LCD	OLED
Response time	Slow	Fast
Luminance boost	Difficult	Possible
Viewing angle	Narrower high contrast angle region	Lambertian distribution[a]
Number of components	More	Fewer
Differential aging[b]	Small	Larger
Susceptibility to water and O_2		Larger

a) Outgoing light distribution whose luminance is proportional to $\cos\theta$. To be discussed in Section 4.2.1.4.
b) Luminance reduction in terms of use of a particular pixel and between colors. To be discussed in Section 2.5.

Phenomenon 1 is the narrower definition. Current OLED devices, after Tang and Van Slyke's "first OLED paper," utilize exclusively LED-like emission mechanisms, that is, phenomenon 2.

Table 2.1 lists the differences between a liquid crystal display (LCD) and an OLED display. The OLED has a very short response time and is capable of using "punching" (an imaging technique for enhancing the local luminance to emphasize the highlighted region of an image). The punching technique is used in cathode ray tubes (CRTs), which can have much higher luminance of a dot than the screen luminance. An OLED can use a similar operation, while a normal LCD display cannot.

Table 2.2 outlines the chronological history of OLED technology development.

The chronological sequence of development listed in Table 2.2 reflects the emergence of some general terms of classification of OLED technologies, including the following:

- Small-molecule OLED (SMOLED) and polymer OLED (PLED)
- Passive-matrix OLED (PMOLED) and active-matrix OLED (AMOLED) displays
- Fluorescent emission and phosphorescent emission.

The developments listed here and in Table 2.2 indicate that the rapid advances in OLED technologies resulted from extensive experimental trial and error. Each technology is discussed in further detail later in the book.

2.1.2 Luminescent Effects in Nature

There are several kinds of "luminescence" in nature, which can be explained by a mechanism similar to that of an OLED.

Table 2.2 Timeline for OLED Technology Development

Year	Event[a]	Company/Institute
1960–mid–1970s	D OLED crystal molecule, anthracene, etc.[b]	NRC (Canada), RCA
1983	D First observation of electroluminescence from polymer film	National Physical Laboratory
1987	P OLED diode structure paper in *Appl. Phys. Lett.*[b]	Eastman Kodak
1988	P Double heterojunction[c]	Kyushu University
1990	P First PLED paper in *Nature*[b]	Cambridge University
1994	P White OLED demonstration[c]	Yamagata University
1996	P first AMOLED demonstration (QVGA)[b]	TDK
1998	D first phosphorescence OLED[b]	Princeton University
1999	D first passive OLED product	Pioneer
1999	D Color OLED display by white + color filter method[c]	TDK
2001	D 0.72-in. headmount display by AMOLED on silicon[b]	eMagin
2001	D 13-in. SVGA AMOLED prototype[b]	Sony
2001	D 2.1-in. 130-ppi AMOLED prototype[b]	Seiko Epson/CDT
2002	D 15-in. 1280×720 OLED prototype[b]	Eastman Kodak/Sanyo
2002	P Tandem OLED device demonstration[c]	Yamagata University
2003	D digital camera with 2.2-in. AMOLED display[b]	Eastman Kodak
2003	D Tiled 24-in. AMOLED prototype with by 12-in. display[b]	Sony
2003	D 20-in. phosphorescence AMOLED prototype by a-Si backplane[b]	ChiMei/IDT/IBM
2006	P White OLED with phosphorescent emitter[c]	UDC
2007	D 11-in. OLED Television product[c]	Sony
2007	P White OLED by all phosphorescent emitters[c]	Konica Minolta
2008	P 12-in. Transparent OLED prototype[c]	Samsung
2008	P 4-in. Flexible OLED prototype[c]	Samsung
2008	P 100% NTSC low power OLED by white + color filter method[c]	Kodak
2009	D OLED lighting product[c]	Philips
2009	P TADF OLED device[c]	Kyushu University

Table 2.2 (Continued)

Year	Event[a]	Company/Institute
2010	P White OLED over 100 lm/W	UDC
2011	D OLED lighting product by all phosphorescent emitters[c]	Konica Minolta/Philips
2013	D 55-in. OLED Television product by white + color filter method[c]	LG display
2013	P 4KOLED Television prototype[c]	Sony/Panasonic
2014	D Flexible OLED display product[c]	LG display
2014	D Flexible OLED lighting product by roll-to-roll manufacturing[c]	Konica Minolta

a) *Abbreviations in this column*: a-si—amorphous silicon; AMOLED—active-matrix OLED; D—development of; P—publication or presentation/demonstration of; PLED—polymer (O)LED; ppi—pixels per inch; QVGA—quarter videographics array (320×240 pixels); SVGA—super videographics array (800×600 pixels).
b) SID International Symposium (2003), 40 Years of SID Symposia—Nurturing Progress in EL/OLED Technology, Baltimore, MD. http://sid.org/Portals/sid/Files/DisplayHistory/EL-OLED_History.pdf.
c) By Takatoshi Tsujimura, "Evolution and future of OLED lighting," OLED Forum Japan presentation, Kyushu University 11/12/2015.

A molecule has multiple discrete energy levels, each able to hold two electrons. When electrons fill these levels completely, beginning from the lowest in energy, the system is stable. This is called the ground state.

If an electron is moved to an upper empty energy level, the resulting configuration is called an excited state. The excited state is normally unstable, so the electron tends to release the energy and return to the ground state. In such a transition, the excess energy is released as light (called "luminescence") and/or heat.

Luminescence can be classified depending on how the energy is transferred to the system. Light emission by a firefly is a type of luminescence. Luciferin is oxidized to "oxyluciferin" in an excited state by luciferase enzyme in the presence of adenosine triphosphate. As the oxyluciferin converts to its ground state, the molecule loses energy that it emits as light. If an excited state created by a chemical reaction causes the photon emission as in this case, it is referred to as chemiluminescence.

Fluorescent paint shines when it is illuminated by short-wavelength light. By absorption of this light, an excited state is created and it causes light emission at a longer wavelength, termed photoluminescence (PL).

When energy is transferred to the molecule by an electric field or current, subsequent light emission is called EL.

There are also many other luminescence effects, such as mechanolumines-cence, and so on. Many of these phenomena have similar mechanisms, in which excess energy is released as photons and the system relaxes to the ground state, as a result.

2.1.3 Difference Between OLED, LED, and Inorganic ELs

OLED, LED, and inorganic EL devices all emit light electrically, so the distinction between them may be confusing. Here, the features for each technology are summarized.

2.1.3.1 Inorganic EL

An inorganic EL device has a structure similar to an OLED, with emissive layers sandwiched by insulators and two electrodes. However, the emission mechanism is quite different.

An inorganic EL device's emissive layer is a mixture of a semiconductor and a metallic compound. Free electrons in the semiconductor are accelerated by an electric field and their kinetic energy is transferred to emission centers, which are excited by their impact. The excited emission center is unstable and loses its energy by emitting a photon.

If a DC voltage were used for electron acceleration, the emission would stop when the electric field in the semiconductor was sufficiently shielded by the electron movement. Therefore, inorganic EL devices are driven by an AC voltage so that the emission is continued.

2.1.3.2 LED

Normally, inorganic LEDs are simply called "LEDs." Though the materials are different, LEDs and organic LEDs have similar emission mechanisms.

Pure semiconductors normally have very few mobile charge carriers. For electronic purposes, doping of impurities into the semiconductor is frequently used to change its conductivity. The impurities may be of two kinds.

1. "Donor"-type impurities, which supply electrons to semiconductors. Electrons are the major charge carriers in this case and such semiconductors with donor impurities are called "n-type."
2. "Acceptor"-type impurities, which supply holes to semiconductors. (As a "hole" means "lack of an electron," supplying a hole is equivalent to accepting an electron.) Holes are then the major charge carriers and semiconductors with acceptor impurities are called "p-type."

If layers of n-type and p-type semiconductors are stacked together to form a junction with electrodes sandwiching the p/n junction layers, electrons from the n-type semiconductor and holes from the p-type semiconductor recombine when electric current is applied between the two electrodes. Recombination in the p–n junction creates excited states that emit photons.

Thus, inorganic LEDs have a mechanism similar to that of organic LEDs, which also emit light due to electron–hole recombination. (To be precise, the conduction mechanisms of LEDs and organic LEDs are not the same. Inorganic semiconductor charge carriers are described by a band theory, in which the periodic potential of the semiconductor crystal leads to formation of a Bloch wave. This theory cannot normally be applied to the disordered layers formed by the organic molecules used for OLEDs. So LEDs and OLEDs operate by the same charge recombination route but due to different conduction mechanisms.)

2.2 BASIC DEVICE STRUCTURE

Emission from all OLED devices—whether of the small-molecular or polymer family—can be explained by the same principle. Through electron–hole recombination, a high-energy molecular state is formed. This state is called an *exciton*, as it behaves like a single molecule with high energy. This exciton emits light after an exciton lifetime period (Figure 2.1). (It should be noted that "exciton lifetime" is the exciton decay period and does not refer to the OLED device lifetime.) [Another type of emission, termed *photoluminescent* (PL) emission, is caused by light (e.g., UV)-induced molecular excitation.]

The wavelength of this light emission corresponds to the exciton energy, so it is possible to control the color of the emission by adjusting the molecular design of the color center. This feature is quite advantageous for OLED display applications.

In experiments using tetracene-doped anthracene crystals and materials, OLED emission had been observed before the so-called first OLED paper in 1987 [5] (see Row 1 in Table 2.2 [4]). However, the operating voltage and efficiency levels were insufficient for actual application. The structure depicted in Figure 2.2 and described in the Tang–Van Slyke paper [5] represents advanced concepts that remain valid today:

1. Significant enhancement of the recombination efficiency by using a layered structure using multiple different materials (heterostructure)
2. Fabrication of low-voltage, high-quality devices through evaporation
3. Appropriate choice of electron and hole injection materials and of work-functions for cathode/anode electrodes
4. High electric field obtained by ultrathin-film formation.

OLED devices could emit very dim light before these developments, but high-luminance operation was achieved only after the first OLED paper.

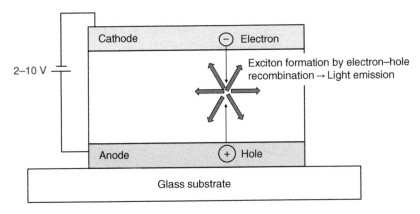

Figure 2.1 Diagram of the OLED emission mechanism.

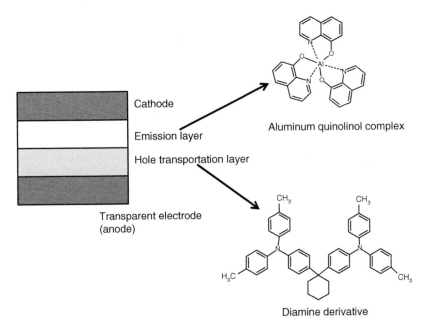

Figure 2.2 OLED device reported by Tang and Van Slyke in 1987 [5].

As mentioned earlier, the two basic families of OLED are small molecule and polymer (SMOLED and PLED). In SMOLED devices, a small molecule is deposited by evaporation technique, so the molecular size is small (however, the mass number of a small molecule can be relatively large). On the other hand, many PLED materials have structures containing chemical groups connected together, including structures that promote solubility and others that

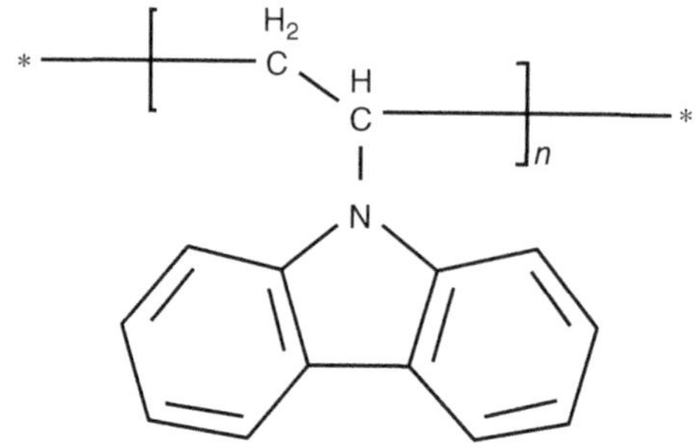

Figure 2.3 Example of a polymer OLED with polyvinylcarbazole molecular structure.

support light emission, the structure having a large overall molecular mass. Figure 2.2 shows typical examples of small-molecule OLED materials: Alq_3 and a diamine derivative. Figure 2.3 presents an example of a polymer OLED material: polyvinylcarbazole (PVK).

2.3 BASIC LIGHT EMISSION MECHANISM

The emission mechanism of OLED is discussed in this section.

2.3.1 Potential Energy of Molecules

When a molecule absorbs light, energy corresponding to the wavelength of light brings the molecule to metastable excited state through an electronic transition. For simplicity, a two-atom molecule, A–B, is assumed. When the distance between the two atoms is r_{A-B}, the potential energy E can be described by Figure 2.4 [6]. The lower curve represents the molecule in its electronic ground state. In the ground state, with a very short r_{A-B}, E increases rapidly. When r_{A-B} is very long, the molecule will become dissociated. The most stable region is at the bottom of the valley in the potential energy curve. The parallel horizontal lines show the energies of different vibrational states. It can also be seen that the potential energy curves of Figure 2.4b also have similar vibrational states.

When a molecule absorbs light, according to the Franck–Condon principle, the molecule increases its energy without changing the interatomic distance. Therefore, the energy increase can be shown as a vertical arrow in Figure 2.4 (called a "vertical transition").

If the potential curve of the excited state does not have a minimum (Figure 2.4a), the excited state is unstable and will dissociate by the increase of r_{A-B} with a corresponding decrease of potential energy. (This situation is called "Predissociation.") In such cases, a continuous optical emission spectrum is observed.

When the excited state potential curve has a valley shape as in Figure 2.4b, the excited state is not easily dissociated; a metastable excited state is formed.

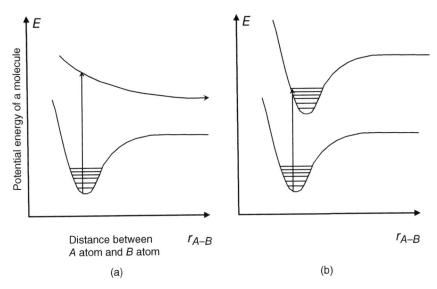

Figure 2.4 Potential curve of ground state and excited state in the case of a two-atom molecule.

As shown in Figure 2.5, several peaks often appear in the absorption spectrum, which correspond to different vibrational levels. (As shown in the figure, the shape of the fluorescent emission spectrum tends to resemble a mirror image of the absorption spectrum. This is because the vibrational level of the excited state and the ground state have similar structures.)

Immediately after light absorption, the interatomic distance in the excited molecule is the same as in the ground state (vertical transition). In a short time, it loses some vibration energy; atoms change their relative positions and the molecule relaxes to the lowest potential energy point along the excited state curve (called "internal conversion," and discussed in Section 2.4.2.1).

2.3.2 Highest Occupied and Lowest Unoccupied Molecular Orbitals (HOMO and LUMO)

When a ground state molecule absorbs light, the molecule changes to an excited state through electron reconfiguration. Using molecular orbital theory, a single covalent bond can be expressed as a linear combination of two electronic wavefunctions. In the case of a single bond, there is a bonding σ orbital with lower energy and an antibonding σ^* orbital with higher energy.

In the case of a double bond, the available energy levels can be shown as Figure 2.6, with σ and σ^* as bonding and antibonding orbitals, respectively, and also π and π^* bonding and antibonding orbitals, respectively.

Figure 2.7 shows the electron configuration of a more complicated molecule in the molecular orbital formalism.

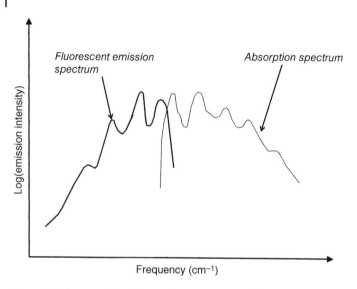

Figure 2.5 An example of absorption spectrum and fluorescent emission spectrum.

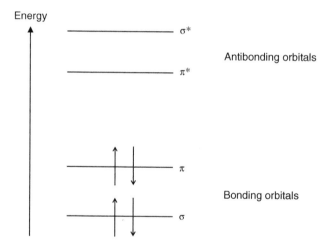

Figure 2.6 Electron configuration of double bond molecular orbital in molecular orbital.

Of all the electron-filled orbitals, the orbital having the maximum energy is called the *highest occupied molecular orbital* (HOMO). Conversely, among the unfilled electron orbitals, the orbital with the lowest electron energy is termed the *lowest unoccupied molecular orbital* (LUMO). The absolute values of the HOMO and LUMO energies relate to the ionization potential and electron affinity of the molecule. The ionization potential is the minimum energy

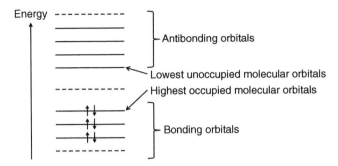

Figure 2.7 Electron configuration in molecular orbital.

required to extract one electron from the HOMO, and electron affinity is the energy released when one electron is added to the LUMO [7].

2.3.3 Configuration of Two Electrons

Before considering the OLED light emission mechanism, it is important for readers to understand the electron configuration in both the ground state and the excited state.

Let us assume that two electrons, 1 and 2, are located in different orbitals. Also, let us define H_1^0 and H_2^0 as Hamiltonians when electrons 1 and 2 exist without interacting:

$$H_1^0 = -\frac{\hbar^2}{2\mu}\nabla_1^2 - \frac{Ze^2}{r_1} \tag{2.1}$$

$$H_2^0 = -\frac{\hbar^2}{2\mu}\nabla_2^2 - \frac{Ze^2}{r_2} \tag{2.2}$$

$$H' = \frac{e^2}{r_{12}} \tag{2.3}$$

Here, H' is a perturbation due to interelectron repulsion.

When two electrons are allocated different orbitals, the Hamiltonian can be expressed as follows:

$$H = H_1^0 + H_2^0 + H' \tag{2.4}$$

If there is no perturbation, the solution of the wave equation can be expressed as follows:

$$\psi = C_1 \chi_A(1)\chi_B(2) + C_2 \chi_B(1)\chi_A(2) \tag{2.5}$$

The calculus of variations is useful here for solving the equation that accounts for perturbation as follows:

$$E\psi = (E^0 + E')\psi = H\psi \tag{2.6}$$

The energy E should be the minimized by the variation:

$$E = \frac{\int \psi H \psi \, d\tau}{\int \psi^2 \, d\tau} \tag{2.7}$$

As ψ_A, ψ_B are orthogonalized and normalized with respect to each other, the equation can be expressed as

$$\int \psi H \psi \, d\tau = \iint \{C_1 \chi_A(1)\chi_B(2) + C_2 \chi_B(1)\chi_A(2)\}(H_1^0 + H_2^0 + H')$$
$$\cdot \{(C_1 \chi_A(1)\chi_B(2) + C_2 \chi(1)\chi_A(2)\}d\tau_1 \, d\tau_2 \tag{2.8}$$

Here

$$\iint \chi_A(1)\chi_B(2)(H_1^0 + H_2^0 + H')\chi_A(1)\chi_B(2)d\tau_1 \, d\tau_2$$
$$= \int \chi_A(1)H_1^0\chi_A(1)d\tau_1 + \int \chi_B(2)H_2^0 \times \chi_B(2)d\tau_2 + J$$
$$= E_A + E_B + J$$
$$= E^0 + J \tag{2.9}$$

and by applying the orthogonalization condition, we obtain

$$\int \chi_A(1)\chi_B(1)d\tau_1 = 0$$
$$\int \chi_A(2)\chi_B(2)d\tau_2 = 0 \tag{2.10}$$

Therefore, Eq. (2.8) can be expressed as follows:

$$\int \psi H \psi \, d\tau = (C_1^2 + C_2^2)(E^0 + J) + 2C_1 C_2 K \tag{2.11}$$

Here, J is the Coulomb integral and K is the exchange integral, which can be expressed as follows:

$$J = \iint \chi_A(1)\chi_B(2)H' \chi_A(1)\chi_B(2)d\tau_1 \, d\tau_2$$
$$= \iint \chi_B(1)\chi_A(2)H' \chi_B(1)\chi_A(2)d\tau_1 \, d\tau_2 \tag{2.12}$$

$$K = \iint \chi_A(1)\chi_B(2)H' \chi_B(1)\chi_A(2)d\tau_1 \, d\tau_2 \tag{2.13}$$

Using Eq. (2.7), we obtain

$$E = \frac{\displaystyle\int \psi H \psi \, d\tau}{\displaystyle\iint \{C_1 \chi_A(1)\chi_B(2) + C_2 \chi_B(1)\chi_A(2)\}^2 d\tau_1 \, d\tau_2}$$

$$= E^0 + H + \frac{2C_1 C_2}{C_1^2 + C_2^2} K \tag{2.14}$$

According to the variational method, a condition for determining the minimum E value needs to be calculated.

A *minimum condition* occurs when the value of differentiation of E by C_1 and C_2 is zero:

$$\frac{\partial E}{\partial C_1} = \frac{\{2C_2(C_1^2 + C_2^2) - 2C_1 C_2 \cdot 2C_1\}K}{(C_1^2 + C_2^2)^2} = 0 \tag{2.15}$$

Therefore,

$$\frac{2C_2(C_2^2 - C_1^2)K}{(C_1^2 + C_2^2)^2} = 0 \tag{2.16}$$

Then

$$C_2 = \pm C_1 \tag{2.17}$$

On the other hand, by applying the normalization condition, we obtain

$$\int \psi^2 d\tau = C_1^2 + C_2^2 = 1 \tag{2.18}$$

Therefore,

$$C_1 = \frac{1}{\sqrt{2}}, \quad C_2 = \pm \frac{1}{\sqrt{2}} \tag{2.19}$$

Using Eq. (2.14), we have

$$E' = E - E^0 = J \pm K$$

Also, by applying Eq. (2.5), we can obtain the following two wavefunctions:

$$\psi_s = \frac{1}{\sqrt{2}}\{\chi_A(1)\chi_B(2) + \chi_B(1)\chi_A(2)\} \tag{2.20}$$

$$\psi_a = \frac{1}{\sqrt{2}}\{\chi_A(1)\chi_B(2) - \chi_B(1)\chi_A(2)\} \tag{2.21}$$

The spin contribution to the molecular system must be considered in addition to the electron configuration determined in Section 2.3.3.

The wavefunction corresponding to a spin quantum number $\pm\frac{1}{2}$ (spin function) is expressed as α or β.

According to the same reasoning applied in Section 2.3.3, the characteristic spin function of a two-electron system can be expressed as follows:

$$\alpha(1)\alpha(2)$$

$$\beta(1)\beta(2)$$

$$\frac{1}{\sqrt{2}}\{\alpha(1)\beta(2) + \beta(1)\alpha(2)\}$$

$$\frac{1}{\sqrt{2}}\{\alpha(1)\beta(2) - \beta(1)\alpha(2)\} \tag{2.22}$$

When the spin function is taken into account, the wavefunction must be anti-symmetric with respect to electron exchange, as indicated in Eqs. (2.20) and (2.21); the wavefunction can then be expressed as follows:

$$\frac{1}{\sqrt{2}}\{\chi_A(1)\chi_B(2) + \chi_B(1)\chi_A(2)\} \cdot \frac{1}{\sqrt{2}}\{\alpha(1)\beta(2) - \beta(1)\alpha(2)\} \tag{2.23}$$

Three further degenerate states with respect to the wavefunctions can be expressed as follows:

$$\frac{1}{\sqrt{2}}\{\chi_A(1)\chi_B(2) - \chi_B(1)\chi_A(2)\}\alpha(1)\alpha(2)\}$$

$$\frac{1}{\sqrt{2}}\{\chi_A(1)\chi_B(2) - \chi_B(1)\chi_A(2)\}\beta(1)\beta(2)\}$$

$$\frac{1}{\sqrt{2}}\{\chi_A(1)\chi_B(2) - \chi_B(1)\chi_A(2)\} \cdot \frac{1}{\sqrt{2}}\{\alpha(1)\beta(2) + \beta(1)\alpha(2)\} \tag{2.24}$$

In total, four states, one in Eq. (2.23) and three in Eq. (2.24), are formed. Equation (2.23) represents *singlet state* and Eq. (2.24), the *triplet state*.

When *2n* electrons are associated with the bonds in a molecule, *2n* molecular orbitals are available to them. In the ground state, electrons occupy these from the lowest to higher energy orbitals in turn. Two electrons with opposite spin occupy each filled orbital as shown in the leftmost part of Figure 2.8. (This is the same case as Figure 2.7.) Normal molecules (those which have no unpaired electrons) have even numbers of electrons and the spins are paired as shown. This state is called S_0, the singlet ground state [6]. When a molecule absorbs light,

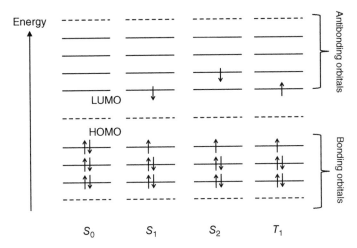

Figure 2.8 Electron configuration in ground state and excited states.

an electron changes its energy level from a bonding orbital to an antibonding orbital. When the electron changes its energy level to the lowest vacant orbital without changing its spin, as shown in the second left part of Figure 2.8, it is called S_1, the first excited singlet state. When a molecule is excited by a higher energy photon, that is, shorter wavelength light, the molecule can be changed to a higher energy state such as $S_2, S_3, S_4, \ldots$.

If the electron changes its spin state on absorbing the photon energy, a triplet state results. If it changes its energy level to lowest vacant orbital and changes its spin, the resulting state is called T_1, the lowest excited triplet state, as shown in the rightmost part of Figure 2.8.

In reality, Figure 2.8 presents a simplified description. The excitation of electron(s) causes the molecule to change its configuration so that it becomes more stable in terms of energy, which also changes the energy levels of orbitals. Figure 2.9 shows the situation for an ethylene molecule. The energy gap is reduced by excitation. (This is why the energy difference between LUMO and HOMO is not in general the same as the energy difference between S_1 and S_0.)

Figure 2.10a shows the energy level of different molecular states. (This kind of energy level diagram of molecular states is called a Jablonski diagram. Jablonski diagrams are very useful to describe molecular state changes. It should be noted that the energy level in a Jablonski diagram shows the energy of molecular states. On the other hand, the energy levels in Figure 2.8 are those of orbitals.) As shown, triplet state T_1 has a lower energy than singlet state S_1. This cannot be explained by simple classical mechanics but requires a quantum mechanical discussion. Any two electrons of the same spin cannot occupy the same orbital, according to Pauli's exclusion principle, while two electrons with different spin states can exist in the same orbital. In a triplet state, where two electrons have

Energy

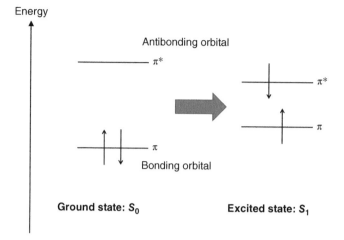

Antibonding orbital

π^*

π

Bonding orbital

Ground state: S_0   **Excited state: S_1**

Figure 2.9 Electron configuration of ethylene molecule in S_0 ground state and in S_1 excited state.

(a) Energy

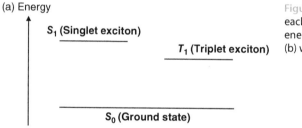

S_1 (Singlet exciton)

T_1 (Triplet exciton)

S_0 (Ground state)

Figure 2.10 Energy state of each exciton (Jablonski energy diagram) (a) with and (b) without vibration states.

(b) Energy

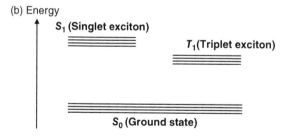

S_1 (Singlet exciton)

T_1 (Triplet exciton)

S_0 (Ground state)

the same spin, the possibility of having the two electrons in the same location is zero. A quantum mechanical calculation also shows that the possibility of having the two electrons in the same vicinity is very low. In a singlet state where the electrons have different spins, the possibility of having them both in the same vicinity is much larger than in the case of a triplet. The difference in triplet

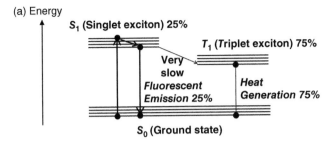

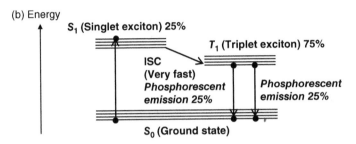

Figure 2.11 Emission mechanisms of (a) fluorescence (singlet) and (b) phosphorescence (triplet) (Jablonski energy diagram).

and singlet wavefunction distributions results in a lower repulsive energy in the triplet state compared to the singlet.

In a real molecule, there are various states due to the excitation of vibrations in the chemical bonds. Figure 2.10b shows the energy diagram taking vibrational states into account.

The excited molecules (excitons) lose their excess energy by one of the following routes (Figure 2.11):

(1) Energy loss by fluorescent light emission
(2) Energy loss by phosphorescent light emission
(3) Energy loss as heat by internal conversion (see Section 2.4.2.1)
(4) Energy causing a chemical reaction to form different molecules.

Regarding (1) and (2), light emission resulting from transition from a singlet state to the ground state is termed *fluorescent emission*; light emission from a triplet state to ground state transition is called *phosphorescent emission*

When two electrons are located in the same orbital, they form an *electron pair* with different spin states. This phenomenon is also known as the Pauli exclusion principle, which states that any two fermions (particles that follow the Fermi–Dirac statistics) of the same type (such as two electrons) cannot occupy the same quantum state simultaneously. Then, the spin direction must be reversed when the electron is changed from the T_1 to the S_0 state.

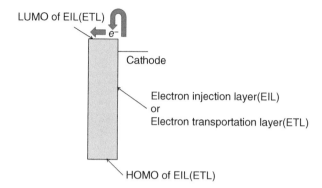

LUMO of EIL(ETL)

e^-

Cathode

Electron injection layer(EIL)
or
Electron transportation layer(ETL)

HOMO of EIL(ETL)

Figure 2.12 Electron injection from cathode to LUMO level of EIL(ETL) material.

Usually, this change is relatively slow, so the exciton lifetime (exciton decay time) of the triplet state is relatively long compared to that of singlet emission. Phosphorescent (triplet) light-emitting materials generally have complex metal structures including heavy metals such as iridium or platinum, as shown in Figure 3.7. These enable energy transfer between singlet and triplet (heavy-metal effect) that is normally prohibited. Due to the heavy-metal effect, both nonradiative transition from S_1 to T_1 (Intersystem crossing, discussed in Section 2.4.2) and radiative transition from T_1 to S_0 are accelerated. As triplet to singlet energy transfer, which is prohibited in the case of fluorescence, is allowed in phosphorescent case, high-efficiency emission is observed.

2.3.6 Charge Injection from Electrodes

Although an OLED device requires recombination of electrons and holes to emit light, normal OLED materials actually have very high resistance under weak electric fields and thus can be regarded as insulators. As discussed in Section 2.2, Tang and Van Slyke's work [5] was so successful because they introduced a very thin film to create strong electric fields and also chose structures and materials suitable for charge injection. State-of-the-art OLED devices also use the same concept. By means of a strong electric field, electrons are injected from a cathode to the LUMO level of the electron transport layer (Figure 2.12) and holes are injected from an anode to the HOMO level of the hole transport layer (HTL) (Figure 2.13).

The current–voltage curve of an OLED normally follows the space-charge-limited current (SCLC) equation (Eq. [2.45]) discussed in Section 2.3.7.2 with the charge injection mechanism discussed in this section.

To understand the mechanism of charge injection into an OLED device, it is very important to understand Schottky thermionic emission[8], tunneling injection[9, 10], and the vacuum-level shift model.

Figure 2.13 Hole injection from anode to HOMO level of HIL material.

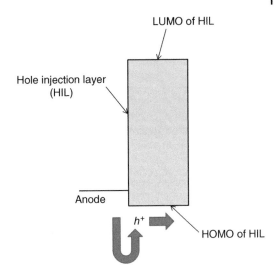

LUMO of HIL

Hole injection layer (HIL)

Anode

h^+

HOMO of HIL

2.3.6.1 Charge Injection by Schottky Thermionic Emission

A thermal electron emerges from the surface of a metal or semiconductor which is heated. The electrons require sufficient energy to cross the potential barrier to exit the electrode surface. When the temperature is increased, electrons possessing sufficient kinetic energy can reach and then exit the surface. This phenomenon is called *thermionic emission* or the *Richardson effect*.

The number of electrons with energy ranging between E and $E+dE$ that reach the metal surface can be expressed as

$$\Delta n(E)dE = 2v_{e,x}N(E)f(E)dE \tag{2.25}$$

where $v_{e,x}$ is the velocity component perpendicular to the surface where the electrons emerge, $N(E)$ the density state, and $f(E)$ the Fermi–Dirac distribution function of electron energy.

The total number of electrons that reach the surface per unit area and per unit time is expressed as

$$n = \int \Delta n(E)dE$$
$$= 2\int v_{e,x}N(E)f(E)dE \tag{2.26}$$

The velocity of an electron can be expressed as follows:

$$mv_{e,x} = \hbar k_x = \hbar k \cos\theta$$

Then

$$n = \frac{2}{(2\pi)^3} \int v_{e,x} f(E(\vec{k})) d^3 \vec{k} \tag{2.27}$$

$$= \frac{2}{(2\pi)^3} \int \frac{\hbar k \cos\theta}{m} \cdot \frac{1}{1 + \exp\frac{\hbar^2 k^2/2m - E_F}{k_B T}} k^2 \sin\theta \, dk \, d\theta \, d\varphi \tag{2.28}$$

Electrons emitted thermionically must have sufficient kinetic energy orthogonal to the surface to penetrate the energy barrier. Therefore, the condition

$$\frac{(\hbar k_x)^2}{2m} = \frac{(\hbar k \cos\theta)^2}{2m} > E_F + \varphi \tag{2.29}$$

must be satisfied for the electrons to be emitted.

The lower bound of the integral in Eq. (2.28) should be

$$k_0 = \sqrt{\frac{E_F + \varphi}{\hbar^2 \cos^2\theta/2m}}$$

Then

$$J = \frac{2e}{(2\pi)^3} \int_0^{2\pi} \int_0^{\pi/2} \int_0^\infty \frac{\hbar k \cos\theta}{m}$$
$$\cdot \frac{1}{a + \exp\left(\frac{\hbar^2 k^2/2m - E_F}{k_B T}\right)} k^2 \sin\theta \, dk \, d\theta \, d\varphi \tag{2.30}$$

At sufficiently high temperatures, the Fermi–Dirac function follows the Maxwell–Boltzmann distribution function approximately. The Maxwell–Boltzmann distribution function can be expressed as

$$g(E) = \exp\left(-\frac{\hbar^2 k^2/2m - E_F}{k_B T}\right) \tag{2.31}$$

so the electric current density of thermionic emission is

$$I = A_0 T^2 \exp\left(-\frac{\varphi}{k_B T}\right) \tag{2.32}$$

This is the *Richardson–Dushman equation*, where I is the saturated emission current density and A_0 is the Richardson constant, which can be expressed as follows:

$$A_0 = \frac{emk_B^2}{2\pi^2\hbar^3} = 120.4 \text{ A/cm}^2 \text{ K}^2 \tag{2.33}$$

In a practical electrode, the injection barrier will be lowered according to the image force principle. This is termed the *Schottky effect* or *barrier lowering* (Figure 2.14).

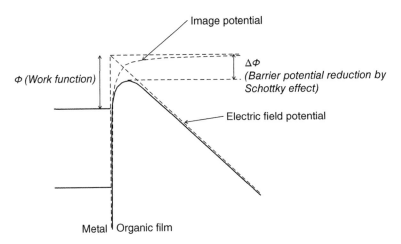

Figure 2.14 Barrier lowering due to image potential.

According to the Schottky effect, the apparent workfunction [11] is lowered by a quantity

$$\Delta\varphi = \sqrt{\frac{e^3 V}{4\pi\varepsilon_0}} \tag{2.34}$$

Therefore, the measured electric current value J can be expressed, using I in Eq. (2.32), as follows:

$$J = I\exp\frac{\Delta\varphi}{k_B T}$$

$$= I\exp\sqrt{\frac{e^3}{4\pi\varepsilon_0}}\cdot\frac{\sqrt{V}}{T} \tag{2.35}$$

In a plot with $\sqrt{V}$ on the abscissa and J on the ordinate, the intercept of the ordinate displays the saturated emission current I. This format is known as a *Schottky plot*.

At the same time, by using the natural logarithm of Eq. (2.32), we can obtain

$$\ln\frac{I}{T^2} = -\frac{\varphi}{k_B}\cdot\frac{1}{T} + \ln A_0 \tag{2.36}$$

Assuming the slope of the graph to be a, with $1/T$ on the abscissa and the logarithm of I/T^2 on the ordinate, we can express the workfunction as

$$\varphi = a\cdot\frac{k_B}{e}eV$$

where k_B denotes the Boltzmann constant and e is the elementary unit charge. In graph format, this relationship is termed the *Richardson plot*.

In an actual OLED device, calculation using this method sometimes yields an apparent energy barrier lower than the theoretical value. Analysis of injection due to states altered by defects or impurities can explain this phenomenon.

2.3.6.2 Tunneling Injection

Some charge injection cannot be analyzed using the Schottky plot and can be better explained by tunneling injection.

Tunneling injection can be expressed by factoring in the Fowler–Nordheim current as

$$J = \frac{q^3 V^2 m_0}{8\pi h \Phi_{Bn} m^*} \exp\left(-\frac{4\sqrt{2m^*}\Phi_{Bn}^{1.5}}{3hqV}\right) \tag{2.37}$$

where m_0 is the free electron mass and m^* is the effective mass number. It should be noted that Eq. (2.37) contains no temperature contribution.

2.3.6.3 Vacuum-Level Shift

The vacuum-level shift model is often used to account for the electron injection mechanism to organic materials from a LiF/Al cathode, which is frequently used as an electron injection layer in OLED devices.

A metal/organic interface or metal halide/organic interface cannot be adequately explained by a Mott–Schottky model, which describes the Fermi energy level alignment by band bending. In the OLED device case, the vacuum-level shift caused by interface electric double layer formation [12–15] is dominant. Sometimes the shift can be as large as 1 eV or more, which cannot be disregarded.

Figure 2.15 shows the band diagram with and without vacuum-level shift [16]. The injection barrier height is lowered due to the electric double layer formed by the LiF/organic interface.

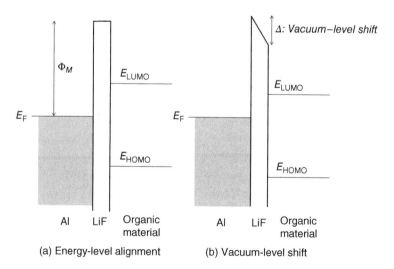

Figure 2.15 Band diagram with and without vacuum-level shift.

Hence, LiF significantly improves electron injection from Al to organic materials, so it is one of the most popularly used cathode electrodes.

Several explanations have been proposed of the vacuum-level shift between an electrode and an organic material [17]:

1. Polarization of the interface due to creation of chemical bonding or charge transfer
2. Polarization due to a mirror potential near the electrode
3. Reduction of the electric double layer due to push back of electrons spilling out into vacuum by Coulomb repulsion of organic molecules
4. Creation of an electric double layer due to charge exchange between interfacial states and the metal
5. Formation of an interfacial electric double layer due to the alignment of organic molecules having a permanent dipole.

Many models have been proposed. As the vacuum-level shift plays a very important role in OLED charge injection, a full understanding of its mechanism will lead to an improvement in OLED device performance.

2.3.7 Charge Transfer and Recombination

2.3.7.1 Charge Transfer Behavior

When a charge is injected into an electrode, if all the HOMO orbitals are occupied and cannot accept additional charge, the charge will be transferred into a LUMO in the case of an electron from a cathode. The electrons transferred from the cathode to LUMOs form an electric current (Figure 2.16). Similarly, holes are injected from the anode and transferred to HOMOs to form a hole current. (Figure 2.17. As a "hole" means the absence of an electron, this eventually means electron removal from an organic material and its transfer to the anode. This HOMO–LUMO charge transfer resembles charge transfer by conduction and valence bands in semiconductors.)

The mobility in OLED thin films ranges from about 10^{-3} to $\sim 10^{-6}$ cm/V·s, and as $\pi-\pi$ interaction is not significant in amorphous films, it is believed that the charges are transferred by hopping conduction, which often follows the space-charge-limited-current rule.

2.3.7.2 Space-Charge-Limited Current

As charge density in the organic material is normally small, when the amount of injected charge exceeds the internal charge amount, the conduction changes from ohmic to what is called "SCLC." To explain this term, when current flows through a vacuum between a cathode and an anode, the electrons carrying it form a space charge. This causes an alteration of the electric field near the cathode and limits the current that can flow. In an organic semiconductor, similar

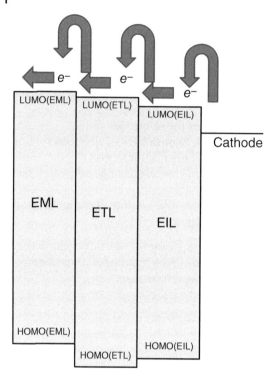

Figure 2.16 Electron transfer in EIL/ETL/EML.

effects occur. Charge carriers cause an electric polarization of the material, which behaves as a leaky dielectric and changes the internal field. Charge carriers can migrate significant distances through the dielectric but become trapped and cannot discharge at an electrode. Similar effects occur in an organic semiconductor. In other words, when charges flow in organic films, local polarization occurs in the organic materials. If charges are injected into the organic film before the polarization is relaxed, a polarized layer is created near the electrode. This is called the *space-charge* layer. Current flow is restricted by this space-charge layer, so it is called SCLC [18].

If charge transfer under an electric field (drift current) and by diffusion (diffusion current) is taken into account, and no charge trapping is assumed, the electric current can be expressed as

$$I = ne\mu E + eD\frac{dn}{dx} \tag{2.38}$$

where n is the charge density injected from the electrode into a thin film of thickness d. This follows the one-dimensional assumption, according to which the x axis is along the thickness direction, μ is the charge mobility, and D is the diffusion constant.

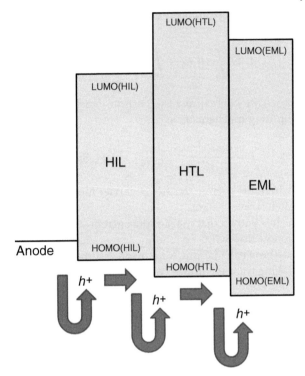

Figure 2.17 Hole transfer in HIL/HTL/EML.

Using the Poisson equation, we obtain

$$\frac{dE}{dx} = -\frac{en}{\varepsilon} \tag{2.39}$$

If an electric field E is applied to the material and if the drift current is sufficiently large compared to the diffusion current, the diffusion term can be neglected. Then, according to Eqs. (2.38) and (2.39), we obtain

$$I = -\varepsilon\mu E\frac{dE}{dx} \tag{2.40}$$

By integration, we then obtain

$$Ix = -\frac{\varepsilon\mu E^2}{2} + C \tag{2.41}$$

If $E = 0$ at $x = d$, then

$$E = \sqrt{\frac{2I(d-x)}{\varepsilon\mu}} \tag{2.42}$$

If the applied voltage is V, we obtain

$$V = \int_0^d E\, dx = \int_0^d \sqrt{\frac{2I(d-x)}{\varepsilon\mu}}\, dx = \frac{2}{3}\sqrt{\frac{2I}{\varepsilon\mu}}d^{3/2} \tag{2.43}$$

Then we can express the current density J, by factoring in the space-charge-limiting mechanism, as

$$J = \frac{9}{8}\frac{\varepsilon\mu V^2}{d^3} \qquad \text{(2.44, Space Charge Limited Current equation)}$$

where J is current density, μ carrier mobility, d thickness, V voltage, and ε the dielectric constant.

It is known that most of the current–voltage characteristics of OLED devices follow this SCLC equation.

These calculations were made assuming no charge trapping. In actual cases, taking trapping into account, the current density is expressed as

$$J = \frac{9}{8}\frac{\varepsilon\mu V^2}{d^3}\cdot\theta$$
$$\theta = \frac{N_c}{N_t}\exp\frac{E_t - E_c}{k_B T} \tag{2.45}$$

where N_c is the effective state density at the conduction band E_c and N_t is the state density of trapping state E_t.

According to Eq. (2.45), the electric current is proportional to the square of the voltage [19]. It should be noted that the current is not related to the carrier density but is proportional to the carrier mobility in this mechanism.

2.3.7.3 Poole–Frenkel conduction

Although Eq. (2.45) describes the behavior of most OLED devices, it was found that some systems such as ITO/PPV/Au do not obey it, especially in the higher electric field region. To explain the observed phenomena, the Poole–Frenkel conduction model, which has field-dependent mobility, was introduced [17].

Similar to the charge injection due to Schottky thermionic emission discussed in Section 2.3.6.1, the barrier height of the trapping state is lowered by

$$\Delta\phi_{PF} = \sqrt{\frac{qE}{4\pi\varepsilon\varepsilon_0}} \tag{2.46}$$

[20], where E is electric field.

Then the mobility can be expressed as follows:

$$\mu(E) = \mu_0 \exp\left(-\frac{E_a - q\Delta\phi_{PF}}{kT}\right)$$

$$= \mu_0 \exp\left(-\frac{E_a}{kT}\right) \cdot \exp\left(\frac{q}{kT}\beta\sqrt{E}\right)$$

$$= \mu_1 \cdot \exp\left(\frac{q}{kT}\beta\sqrt{E}\right) \tag{2.47}$$

where μ_1 is the mobility at zero electric field and can be expressed as

$$\mu_1 \equiv \mu_0 \exp\left(-\frac{E_a}{kT}\right) \tag{2.48}$$

and β is the Poole–Frenkel factor and is expressed as

$$\beta = \sqrt{\frac{e}{\pi\varepsilon\varepsilon_0}} \tag{2.49}$$

This is called the Poole–Frenkel conduction model.

The SCLC in Eq. (2.45) can be rewritten as

$$J(E) = \frac{9}{8}\varepsilon\frac{E^2}{d}\mu(0)\exp(\beta\sqrt{E}) \tag{2.50}$$

(SCLC with field-dependent mobility).

The actual situation is more complicated; the energy and location of trapping states have distributions, not a single value. There are several models to account for such cases. Among them, the "multiple trapping and release" (MTR) model is the most popularly used and can explain many of the phenomena. The MTR model assumes a narrow delocalized band with a high concentration of localized trapping levels [21]. During charge movement through the delocalized states, the charge is trapped and is then thermally released. As a result, the drift mobility μ_D due to the MTR mechanism can be expressed as

$$\mu_D = \mu_0\alpha\exp\left(-\frac{E_t}{kT}\right) \tag{2.51}$$

α is the ratio of density of states at the band edge to the trap concentration, in the case of a single trapping level. In the case of an energy-distributed trap, the drift mobility can be expressed by the same equation as Eq. (8.5) with E_t and α calculated by parameter fitting. (The MTR model can be also used for amorphous silicon TFTs.)

2.3.7.4 Recombination and Generation of Excitons

As shown in Figure 2.18, injected electrons and holes are transferred through their own charge transport layers in an OLED device. When an electron meets a hole, the electron and hole recombine and an exciton is formed. An exciton generates a photon whose wavelength corresponds to the energy gap after

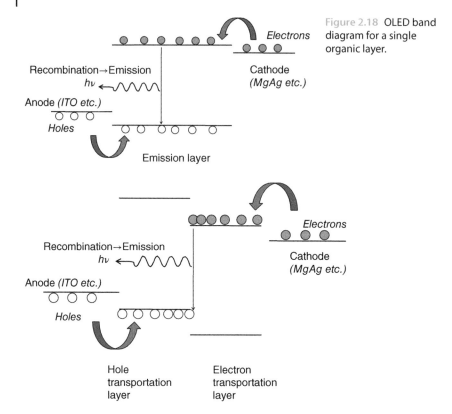

Figure 2.18 OLED band diagram for a single organic layer.

Figure 2.19 Enhancement of electron–hole recombination efficiency by multilayer structural configuration.

completion of its exciton lifetime or loses its energy due to internal conversion discussed in Section 2.4.2. If the probability of electron–hole recombination can be increased, more photons will be generated. So if electron and hole transport can be blocked (i.e., if their flow can be impeded) by the HOMO/LUMO band energy change at an interface between different material layers as shown in Figure 2.19, the location of their recombination can be controlled. As the charge carriers cannot easily pass through the entire device to reach the electrode [5], the recombination probability can be enhanced, and high efficiency can be achieved.

Recently developed OLED devices normally use structures that apply advanced modifications to this charge blocking principle to achieve efficient carrier recombination (Figure 2.20).

For example, the following are typical layers for an OLED device, using material suitable for each function:

- Electron injection layer (EIL)—for electron injection from the cathode

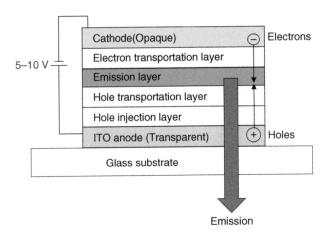

Figure 2.20 Example layout of an OLED device using role sharing by multiple layers.

- Electron transport layer (ETL)—for electron injection from the cathode and transport of electrons
- Hole blocking layer (HBL)—for blockage of hole transport
- Emission layer (EML)—electron/hole transport and their recombination to form an exciton, which generates light emission
- HTL—for hole transport from HIL to EML layer
- Hole injection layer (HIL)—for hole injection from the anode.

EIL and HBL can also be used to control the location where the excitons are formed by blocking hole transportation.

The choice of electrodes is also very important. For example, the HOMO level of NPB is −5.5 eV, and the workfunction of indium tin oxide (ITO) is about −4.7 eV. By solution modification, the workfunction of ITO can be changed to around −5.0 eV, which improves the hole injection efficiency; placement of a thin layer of CuPc (copper phthalocyanine) between ITO and NPB also helps improve the injection efficiency. (Regarding the anode preparation, see Section 6.2.2 for detail.)

Regarding the cathode side, as the LUMO of Alq_3 is around −3.3 eV, a MgAg alloy with a −3.8 eV workfunction is sometimes used as a cathode. It is also known that the injection efficiency using Al with workfunction −4.3 eV can be improved by placement of an LiF layer between Al and Alq_3. (See vacuum-level shift described in Section 2.3.6.3.)

Therefore, it is very important to design an OLED device by factoring in the relationships between workfunction, HOMO level, and LUMO level of each layer. To estimate such an energy level, ionization potential measurement is important (see the discussion in Section 3.4.1).

Since the 1980s, OLED device design has been significantly improved using the thin-film role sharing described above.

2.4 EMISSION EFFICIENCY

2.4.1 Internal/External Quantum Efficiency

To quantify how effectively an OLED device is emitting light, the internal quantum efficiency (IQE) and external quantum efficiency (EQE) are often calculated. Internal quantum efficiency η_{int} can be expressed by the following equation, which indicates how much light is generated from a given injected charge:

$$\eta_{int} = \gamma\, \zeta_s q \tag{2.52}$$

Here, γ is the electron hole balance (carrier balance), ζ_s the exciton generation efficiency (for singlet emitters, this is the singlet generation efficiency; in the case of triplet emission, this should be the sum of the singlet generation efficiency and the triplet generation efficiency), and q the fluorescence quantum efficiency for singlet (the efficiency to generate fluorescence light emission from the exciton) and phosphorescence quantum efficiency for triplet (the efficiency to generate phosphorescence emission). There are many reasons for an exciton to lose its energy without light emission, as discussed in Section 2.4.2.

In actual devices, a significant proportion of generated light is reflected, transferred, and absorbed in both the active layers and the substrate, so it is necessary to also consider the outcoupling efficiency (ratio of generated to extracted light) in order to understand the luminance when designing a display.

When the refractive index of a substrate is n (*refractive index refers to the ratio of light speed in vacuum divided by the phase velocity of light in the medium*), outcoupling efficiency can be expressed as η_o (refer to Section 2.4.3 for further details).

External quantum efficiency can be expressed as follows:

$$\eta_{int} \cong \eta_o \gamma\, \zeta_s q \tag{2.53}$$

Using the most simplified outcoupling theory using classical reflection and refraction theory, η_o can be written as [21]

$$\eta_0 = 1/(2n^2) \tag{2.54}$$

In the case of fluorescent emission, the maximum value of the carrier balance is 100%, singlet generation efficiency is 25%, the maximum value of fluorescence quantum efficiency is 100%, and outcoupling efficiency is 20%, (n_{glass}=1.6 in Eq. [2.54]) the EQE limitation obtained using the fluorescent mechanism would

Figure 2.21 Explanation of internal conversion by potential curve.

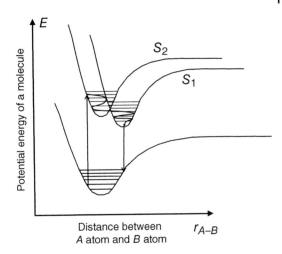

be about 5% (=100% × 25% × 100% × 20%). In actual cases, quantum efficiency numbers higher than this have been observed due to, so-called, *delayed fluorescent emission* as discussed in Section 2.4.2.4.

According to similar calculations, the EQE limit for phosphorescence is thought to be around 20% (without light extraction structures, which are used for lighting, discussed in Section 9.4).

2.4.2 Energy Conversion and Quenching

2.4.2.1 Internal Conversion

There are several reasons for exciton energy loss after an electron–hole pair recombines. Figure 2.21 shows the conversion of an S_2 exciton to an S_1 exciton due to the crossover of potential curves [6]. Excitons in higher energy levels such as S_2 and T_2 have shorter lifetime than S_1 and T_1, so they tend to be converted to S_1 and T_1. At the point where the S_1 potential curve and the S_2 potential curve cross in Figure 2.21, S_1 and S_2 have the same potential energy and the same atom configuration. Therefore, S_2 can be easily converted to S_1 in this situation (called internal conversion; T_2 can be converted to T_1 in a similar way).

2.4.2.2 Intersystem Crossing

An S_1 exciton generated by direct excitation, carrier recombination or by internal conversion from a higher energy state (such as S_2) can be converted to T_1, due to crossing of the potential curves corresponding to the T_1 and S_1 states. (ISC: Intersystem crossing.) As S_1 has a single state and T_1 has three states, 25% of singlet and 75% of triplet exciton states can result at equilibrium (discussed in Sections 2.3.5 and 2.4.2.3). T_1 states normally cannot be converted to the S_0 ground state with emission of light, as a transition with a spin state change

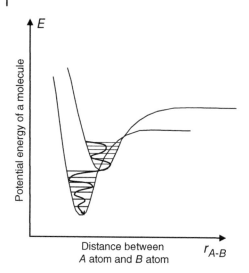

Figure 2.22 Explanation of radiationless deactivation by potential curve.

is forbidden. Thus, fluorescent system has 25% maximum internal quantum efficiency. In this case, the 75% of triplet excitons lose their energy as heat by radiationless deactivation (Figure 2.22). In the phosphorescent system, a phosphorescent emitter doped into the emissive layer converts T_1 to S_0 ground state with light emission; the transition becomes allowed through the heavy-metal effect, which is caused by spin-orbital interaction. Therefore, a phosphorescent system can provide 100% internal quantum efficiency due to emission from both S_1 and T_1 states (discussed in Section 2.4.2.3).

2.4.2.3 Doping

By adding a small quantity of emissive material, it is possible to improve the efficiency or to change the emission color. This is called *doping* [22, 23], and the materials used for doping are called *dopants*. This technique is especially useful for materials in which emission yield decreases at high density (this activity is referred to as *concentration quenching*). For example, in its original state Alq_3 is a green fluorescent emission material, but with the addition of small amount of rubrene, the emission color changes to yellow-orange. This is due to energy transfer from the Alq_3 exciton to the rubrene exciton. With the addition of a red fluorescent dopant to Alq_3 + rubrene system, such as DCJTB (4-(dicyanomethylene)-2-*tert*-butyl-6-(1,1,7,7,-tetramethyljulolidyl-9-enyl)-4*H*-pyran), further energy transfer occurs, and red emission is observed. Thus, it is possible to modify the color by converting the emission spectrum to a longer wavelength.

Figure 2.23 shows the doping emission mechanism in the case of fluorescence [23]. Host energy is transferred to the dopant by dipole–dipole coupling. This

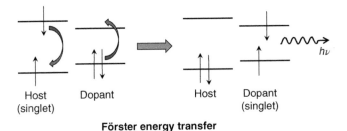

Förster energy transfer

Figure 2.23 Doping emission by fluorescence mechanism.

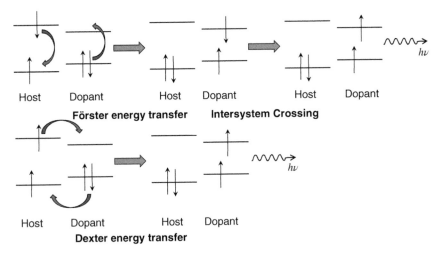

Figure 2.24 Doping emission by phosphorescence mechanism.

energy transfer can occur even when the excitons are separated by relatively long distances such as 100 Å (this is known as *Förster energy transfer*).

When this fluorescent emission mechanism is employed, only 25% of the excitons can be used because only singlet excitons contribute to the emission and the remaining 75% of excitons lose their energy by internal conversion (discussed in Section 2.4.2) to generate heat. For low power consumption and high luminance, emission yield must be improved.

Figure 2.24 shows an example of phosphorescent dopant emission reported by Baldo et al. [24]. They added approximately 6–8% of the phosphorescence dopant to the emissive layer such that both singlet and triplet excitons contributed to the emission. The energy transfer in this procedure [24] differs from Förster transfer, which uses dipole–dipole coupling. In the Dexter energy transfer method described by Baldo's group [24], energy transfer is effected by exciton hopping, in which case a shorter interexciton distance (e.g., ~10 Å) is necessary to make the transfer effective.

2.4.2.4 Quenching

Even if an exciton is in a radiative state, it may still lose its energy without emission if a quencher is present. The quencher can be included when the films are made and also can be created by electrical driving of the device or by aging. The likelihood of quenching tends to increase at higher exciton density (as the excitons have an increased chance to interact or to meet with quenchers, such as increased polaron).

Phosphorescent excitons have an especially long decay time, so are susceptible to energy loss by quenching.

The best-known quencher of phosphorescent emitters is oxygen. Oxygen's ground state is a triplet (the oxygen molecule has a peculiar biradical electron configuration, which has two unpaired electrons with the same spin, as the ground state. It has two electrons with the same state, namely triplet state), so the energy of a triplet exciton can be easily lost.

Other major causes of triplet exciton quenching include the following [25–27]:

1. Triplet–triplet annihilation (TTA)

   Triplet–triplet annihilation (TTA) is a conversion of two triplets into a ground state singlet and an excited singlet [28]. While there are a number of related processes, it has been reported that the process of donor molecule energy exchange, which was caused by two triplet excitons, to an excited acceptor molecule to form singlet state and triplet state according to the following formula, is dominant [29].

$$^3A^* + {}^3A^* \rightarrow (4/9)^1A + (1/9)^1A^* + (13/9)^3A^* \qquad (2.55)$$

   TTA causes efficiency loss at high current density due to the higher chance of triplet–triplet collision at the high triplet exciton densities found in phosphorescent devices. In the case of fluorescent emitters, triplet excitons are normally lost without emission, as the radiative transition from excited triplet to singlet ground state is forbidden. However, many cases have been reported, which exceed the quantum efficiency limit (25%) of fluorescent OLEDs [30, 31]. These observations can be interpreted as an efficiency increase due to TTA. According to (2.55), two triplet excitons (left) generate 1/9 singlet exciton and 13/9 triplet excitons, so in total and assuming the triplet excitons are recycled, five triplet excitons generate one singlet. The maximum internal quantum efficiency that can be achieved using fluorescent emitters then becomes 25% (singlet) plus one fifth of 75% (triplet) = 40%. This increase in emission occurs through TTA—contrary to the usual meaning of the word, so it is sometimes called triplet–triplet fusion (TTF). The singlet excitons generated by TTF decay later than those derived directly from carrier recombination, so delayed emission is observed from this effect. This type of efficiency increase is therefore called delayed fluorescence.

2. Triplet–polaron quenching (TPQ)

   When a triplet exciton interacts with a charged molecule, the energy of a donor molecule can be transferred to the charged molecule.

3. Field-induced quenching [32]

   At a high electric field, the decrease in electrophosphorescence is observed. Also neat solid film also decreased phosphorescence at high electric field. This phenomenon is interpreted that the electron–hole pairs are coulombically dissociated by electric field [33].

   Among these exciton quenching effects, it is reported that only TTA has an important effect on the OLED efficiency [34].

Poor OLED performance and OLED device degradation can sometimes be explained by a quenching mechanism. To analyze the system, it is important to know what will happen if a quencher is added to the OLED emissive system.

When a quencher is added to a photochemical system, the following elementary reactions occur [6]:

1. Photon absorption:

$$A_0 \xrightarrow{hv} A^*$$ (2.56)

2. Photo emission:

$$A^* \xrightarrow{k_e} A_0 + hv$$ (2.57)

3. Quenching:

$$A^* + A \xrightarrow{k_q} A_0 + A^*$$ (2.58)

4. Other deactivation:

$$A^* \xrightarrow{k_d} A_0 + \text{Heat}$$ (2.59)

5. Reaction:

$$A^* \xrightarrow{k_r} P(\text{Product})$$ (2.60)

The rate of each reaction can be written as I_a, $k_e[A^*]$, $k_q[A^*][Q]$, $k_d[A^*]$, and $k_r[A^*]$, respectively.

The rate of excited state generation can be written as

$$\frac{d[A^*]}{dt} = I_a - (k_e + k_q[Q] + k_d + k_r)[A^*]$$ (2.61)

When the system is in equilibrium, left term equals zero.
Therefore,

$$I_a = (k_e + k_q[Q] + k_d + k_r)[A^*]$$ (2.62)

When φ_r is the quantum yield of the reaction and φ_{r0} is the quantum yield without quencher,

$$\varphi_{r0}/\varphi_r = \frac{k_r}{k_e + k_d + k_r} \cdot \left(\frac{k_r}{k_e + k_q[Q] + k_d + k_r}\right)^{-1}$$

$$= 1 + \frac{k_q}{k_e + k_d + k_r}[Q] = 1 + k_q\tau_0[Q] \tag{2.63}$$

where τ_0 is the exciton lifetime of A^*.

When total reaction constant of exciton decay is $k[A^*]$("Photon emission" + "Other deactivation".)

$$\phi_r = \frac{k_r}{k_r + k_d + k_q[Q]} \tag{2.64}$$

Therefore,

$$1/\phi_r = \alpha + \beta \cdot [Q] \tag{2.65}$$

where

$$\alpha = 1 + \frac{k_d}{k_r} \quad \text{and} \quad \beta = \frac{k_q}{k_r}$$

When a photochemical reaction is caused by a single excited state, the graph $1/\varphi r$ versus $[Q]$ shows linear behavior. This graph is called a Stern–Volmer plot.

When photochemical reaction contains multiple phenomena, the Stern–Volmer plot can show nonlinear relationship. For example, at low quencher density, the amount of quenching is proportional to the frequency of collision (dynamic quenching) between an emissive exciton and a quencher, then the Stern–Volmer plot shows linear relationship. However, in the case when the quencher forms nonemissive stable complex (static quenching), Stern–Volmer plot shows superlinear relationship due to two quenching mechanisms. In other cases, for example, when there are singlet and triplet excitons and only triplet excitons are quenched, the Stern–Volmer plot shows saturation at a certain efficiency value, as the quencher volume is increased.

To determine whether poor device performance is due to a quencher, a Stern–Volmer plot analysis is effective.

2.4.3 Outcoupling Efficiency of OLED Display

2.4.3.1 Light Output Distribution
Light generated by excitons travels via various paths. Some of the light reaches the human eye, but a significant portion of the light diminishes by propagation in the thin film (waveguide mode) or by propagation in substrate (substrate mode).

Meerheim et al. made quantitative analysis of energy loss mechanism in OLED [35]. The energy applied is used for the following:

(1) Electrical losses
(2) Nonradiative losses
(3) Absorption
(4) Surface plasmons
(5) Waveguided-mode losses
(6) Substrate-mode losses
(7) Outcoupled light, extracted to outside.

Especially, surface plasmons, waveguided mode, and substrate mode are the three major reasons of losses in OLED devices. It is important to make countermeasure for each light loss mechanism to achieve highly efficient devices. (Light extraction enhancement in the case of OLED lighting is discussed in Section 9.4.)

The basic characteristics of optics are discussed as follows.

2.4.3.2 Snell's Law and Critical Angle

Denoting the incident angle as θ_i the reflection angle as θ_r, and the refraction angle of the transmitted light as θ_t, then according to Snell's law (Figure 2.25), we have

$$\frac{\sin \theta_i}{\sin \theta_t} = \frac{n_2}{n_1} \tag{2.66}$$

where n_1 and n_2 are the refractive indices of materials 1 and 2, respectively.

When θ_i is larger than the critical angle θ_C, the incident light causes total reflection. The critical angle is expressed as follows:

$$\theta_C = \sin^{-1}\frac{n_2}{n_1} \tag{2.67}$$

Figure 2.25 Schematic representation of Snell's law.

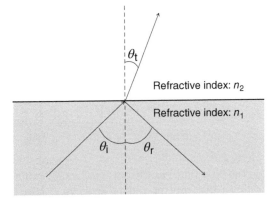

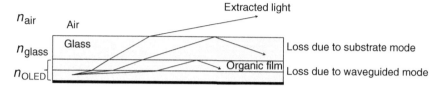

Figure 2.26 Schematic representation of light extraction loss.

If the light in high refractive index keeps its incident angle larger than critical angle, the light never comes out and is lost. This is the major cause of waveguided-mode loss and substrate-mode loss.

2.4.3.3 Loss Due to Light Extraction

For a point light source, the light extraction outcoupling efficiency η_o is

$$\eta_o = \frac{2\pi I_0 \int_0^{\theta_B} \sin\theta\, d\theta}{2\pi I_0 \int_0^{\pi/2} \sin\theta\, d\theta} = 1 - \sqrt{1 - \left(\frac{n_{air}}{n_{OLED}}\right)^2} \tag{2.68}$$

Here, I_0 is the luminous intensity toward the front direction. For a diffusion surface, we obtain

$$\eta_o = \frac{2\pi I_0 \int_0^{\theta_B} \cos\theta\sin\theta\, d\theta}{2\pi I_0 \int_0^{\pi/2} \cos\theta\sin\theta\, d\theta} = \left(\frac{n_{air}}{n_{OLED}}\right)^2 \tag{2.69}$$

For example, if the refractive index of air is $n_{air} = 1.0$ and the OLED organic film refractive index is $n_{OLED} = 1.75$ (Alq$_3$), the light extraction efficiency is very low: approximately 17.9% for the point light source and 32.6% for the diffusion surface (Figure 2.26).

Figure 2.26 is a very simplified hypothetical depiction of light extraction loss, and, in actual cases, the optical properties of multiple films, electrodes, and multiple interference effects must be accounted for. Mikami et al. [36] derived the following equation to determine the energy extracted from the device in this kind of system:

$$E = \frac{(1 - |\rho_0|^2\{1 - |\rho_0|^2 + 2|\rho_0|\exp i(2\delta_1 - \varphi_1)\}}{1 + |\rho_0||\rho_1| - \sqrt{|\rho_0||\rho_1|}\exp i(2\delta_0 - \varphi_0 - \varphi_1)} \tag{2.70}$$

Here, E is extracted energy, ϕ_0 the phase shift due to front surface reflection, ϕ_1 the phase shift due to rear surface reflection, $|\rho_0|^2 = R_0$ the energy reflection ratio of the front boundary layer, $|\rho_1|^2 = R_1$ the energy reflection of the rear boundary layer, Δ_0 the phase shift due to thickness of the EML, and Δ_1 the phase shift that results between the EML and the rear layers.

By adjusting the parameters to ensure uniformity of the light phase, which may be modified by the multilayer reflection, one can utilize the resonance effect (microcavity effect) to achieve high efficiency and color purity in OLED devices. Especially for the top emission case (emission from the top of the structure), the resonance effect can be utilized by using a semitransparent thin metal film for the top electrode. However, the microcavity effect causes significant alteration of characteristics due to thickness variation; thus, precise thickness control is necessary for manufacturing.

In the bottom emission case (i.e., light emission from the bottom of the structure), it is also possible to apply the microcavity effect by using a semitransparent bottom electrode or by employing a Bragg reflector.

Other approaches for enhancement of light extraction have been reported, including microlens arrays, micropyramidal structure, diffraction structure, and photonic arrays. These designs are discussed in further detail in Section 9.4.

2.4.3.4 Performance Enhancement by Molecular Alignment

Since "first OLED paper" was released in 1987, in most cases, the performance of OLED was discussed based on the single molecular behavior. Different from inorganic semiconductors, which has strong interaction between atoms that clearly defines the band structure according to the Bloch theory, the interaction between organic molecules is much weaker due to poor molecular orbital overlap with adjacent molecules. Also it has been proved that amorphous organic film deposition gives large benefit in device fabrication, due to uniformity and prevention of electric shortage between anode and cathode electrodes, so it was perceived as pointless to discuss about the orientation of molecules in the past.

However, it was gradually proved that a sort of orientation in amorphous film is playing an important role in the device performance these days [37].

In general, rod-shaped molecules in amorphous film show horizontal molecular orientation, while molecules in polycrystalline film show vertical orientation [38]. There are two major reasons why horizontal molecular alignment gives better OLED device performance.

1. *Electrical advantage:* Horizontal molecular orientation increases the $\pi-\pi$ overlap between adjacent molecules and decreases the disorder in the film, which improves the charge transportation toward thickness direction.
2. *Outcoupling advantage:* Molecules emit light to perpendicular direction to its dipole moment, so the light from horizontal molecular orientation gives smaller incident angle when it arrives to the film interface. The small incident angle avoids the total reflection caused by the refractive index difference (Figure 2.25). If such horizontally aligned molecules would be fabricated by

some method, it also brings merit to avoid the absorption loss by the circular polarizer (discussed in Section 4.1) because such aligned molecules cause polarized emission.

To make such oriented or aligned molecules happen, several processes have been proposed, such as substrate temperature control during vacuum evaporation [38], rubbing of molecules [39], electric field [40], and stretching [41].

Horizontal alignment of the emitter is also effective to suppress the plasmon-related energy loss. The topic is discussed in Section 9.4.5.1.

2.5 LIFETIME AND IMAGE BURNING

2.5.1 Lifetime Definitions

According to the driving stress, OLED device decays its luminance. The decrease in luminance of the device not only causes the luminance reduction of a display but also causes the luminance difference between the pixels according to the amount of total driving stress applied to the pixels (picture elements). As a result, pixels of a pattern that was displayed for a long period remains as luminance decrease and creates image burning, where black-and-white-reversed image of the displayed pattern remains when the display is operated. This is called differential aging (Figure 2.27).

Figure 2.28 shows the variation in luminance of an OLED device with constant current operation, in which a perylene-doped BAlq layer functions as the EML. As indicated in the graph, the time spent prior to the 50% luminance decay is denoted T_{50} (also as LT_{50}). Similarly, the time spent prior to the 30%

Differential aging

Figure 2.27 Image sticking on a display caused by the differential aging.

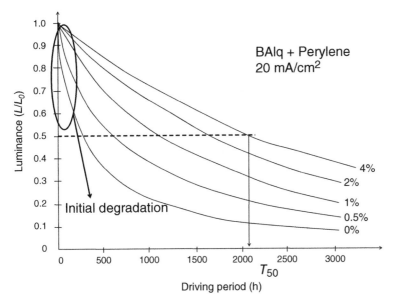

Figure 2.28 Luminance decay curve.

luminance decay is denoted T_{70}; the time prior to 5% decay, T_{95}; and the time prior to 3% decay, T_{97}. Previously T_{50} was used most frequently; however, as OLED lifetime increases every year, and as 50% luminance degradation is seldom acceptable from a product perspective, many OLED display and OLED lighting users now employ T_{70}. Standards T_{95} and T_{97} are also used to show how a display is likely to cause *image burning* because 3–5% display luminance change is close to the perception limit of the human eye. This luminance difference almost corresponds to the color difference $\Delta E = 1$ in uniform color space (discussed in Section 4.2.4). The luminance difference is not associated with color, so this corresponds to $L^* = 1$. The uniform color space concept is very important for determining the allowable limitation as to the luminance decay and color change.

2.5.2 Degradation Analysis and Design Optimization

A closer look at the 0% doping curve in Figure 2.28 reveals a gradual decrease in the rate of degradation from the left to right of the graph [42]. One can also note a sharp drop in the initial decay rate with a decrease in perylene doping. This initial reduction in the decay rate results in total lifetime decrease as well as image burning; these effects can be minimized by fine-tuning the device design and process.

During operation of an OLED display, any color shift should be minimized as well. The pixel is composed of red, green, and blue subpixels as shown in

Figure 2.29. After operation, these red, green, and blue subpixels (spatially located primary color elements in a pixel) are degraded as shown in Figure 2.30. As white is expressed by the mixture (Figure 2.28) of red, green, and blue (the calculation method is discussed in Section 4.2.5.), the white point (color coordinate of white) in the display is shifted on color coordinate if the primary color luminance has been changed.

To avoid the white point shift, the aperture may be adjusted (aperture adjustment, Figure 2.31) according to variation in the degradation curve for individual colors (Figure 2.30).

When the current density is J, the lifetime of an OLED device decreases in proportion to $1/J^n$ ($n = 1.2$–1.9, called the acceleration factor, discussed in Section 2.5.3.1.). By adjustment of the aperture ratio, the current density can be controlled, so the decay curve for each color—red, green, and blue—can be adjusted to a similar curve as shown in Figure 2.32, and white point shift can be reduced.

However, the most recent types of display, such as those for high-resolution mobile phone applications, have design restrictions such as shadow mask intolerance and negligible flexibility for aperture ratio adjustment for each color.

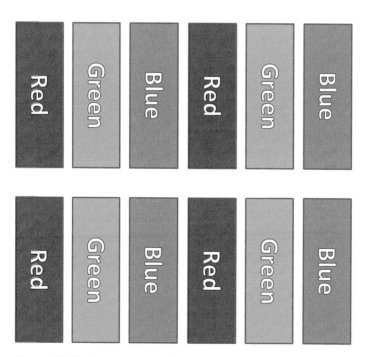

Figure 2.29 Pixel arrangement without aperture adjustment.

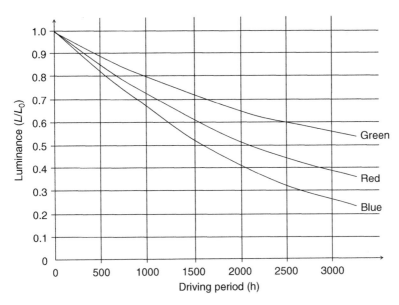

Figure 2.30 Individual degradation curves for three colors.

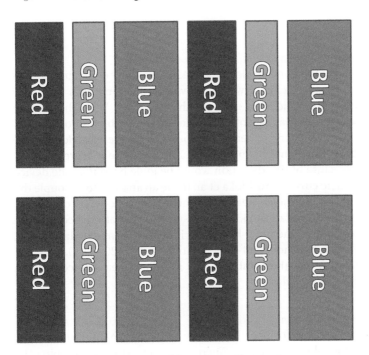

Figure 2.31 Pixel arrangement with aperture adjustment.

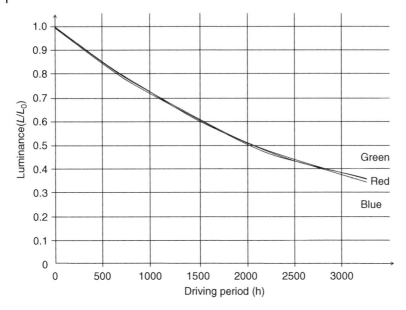

Figure 2.32 OLED pixel display curve for a display with aperture adjustment.

2.5.3 Degradation Measurement and Mechanisms

2.5.3.1 Acceleration Factor and Temperature Contribution

As OLED is a current-driven device, a molecule would be affected by the passing charges. Negative charge would provide reduction stress and positive charge would give oxidation stress to the molecule. With that simple assumption, the electrical damage of OLED device should be proportional to the number of passing charges, hence proportional to the electric current. However, in reality, the damage is proportional to J^n whose n is greater than one. One possible cause of this deviation would be joule heating of the device caused by the electric current. Yoshioka et al. made an analysis to decouple the contribution between non-joule-heat component and joule-heat component of the OLED device degradation [43]. The result shows that the acceleration factor is decreased when the joule-heat contribution is eliminated; however, the acceleration factor without self-joule heating is $n = 1.20 \pm 0.10$, which is not one. This implies that there still remains non-joule-heating acceleration mechanism of device degradation.

2.5.3.2 Degradation Mechanism Variation

For good OLED device performance, it is very important to understand what is happening in the fabricated device. There are many articles describing the cause of OLED device degradation.

Aziz et al. explored the degradation study of OLED device and categorized it into three mechanisms: (1) dark-spot degradation, (2) catastrophic failure, and (3) intrinsic degradation [44].

Burrows et al. made a comparison of the degradation behavior of encapsulated device and nonencapsulated device [45]. They found that the encapsulated device has two orders of magnitude longer lifetime than that of nonencapsulated ones. The encapsulation issue is discussed in Section 3.3.

To understand the cause of encapsulated OLED device degradation, Kondakov et al. made an analysis on the OLED device after 312 h operation at 80 mA/cm^2 stress [46]. The derivatives detected by HPLC and the expected precursors of the detected materials were assumed as quenchers in the study. It was found that the monotonic increase in the extent of NPB chemical degradation coincide with the monotonic decrease in luminous efficacy, which indicates that the nonradiative NPB derivatives created by driving stress cause the quenching of excitons in the system.

Meerheim et al. made an analysis on the hole blocking layer (HBL) dependence of phosphorescent OLED device systems [47]. When BPhen is used as an HBL layer, the OLED device lifetime is much shorter (1/50) in than the case of BAlq HBL. The paper concludes the difference is due to two mechanisms, the BPhen's dimer acting as a quencher and also BPhen acting as a ligand of Iridium, which is causing the nonradiative recombination center.

So et al. made a comparison between small-molecule OLED and polymer OLED in the degradation mechanism [48]. In the small-molecule case, the accumulation of degradation products are causing both luminous efficacy loss and operating voltage increase. In the case of polymer OLED, charge-transport and injection properties affect the device lifetime. It was also pointed out that charge balance plays a very important role in having a good lifetime of the device, as surplus carriers damage the organic materials.

2.6 TECHNOLOGIES TO ENHANCE THE DEVICE PERFORMANCE

2.6.1 Thermally Activated Delayed Fluorescence

As discussed in Section 2.4.2.2, by means of phosphorescent emitter such as Iridium complex, high internal quantum efficiency near 100% is possible, which enables conversion between singlet and triplet by means of intersystem crossing. However, the triplet excitons by phosphorescent emitters have several issues such as triplet exciton quenching (Section 2.4.2.4) and high cost due to Iridium compound, so new method to achieve high quantum efficiency has been suggested.

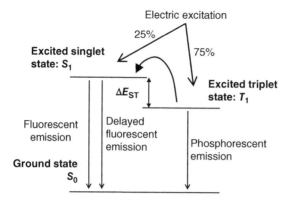

Figure 2.33 Energy diagram of TADF system.

Figure 2.33 shows the energy diagram to explain the light emission mechanism from S_1 and T_1 states. In fluorescent system, phosphorescent emission from T_1 to S_0 is forbidden, so 75% of excitons in the triplet state cannot contribute to the emission normally. To allow such a triplet state to emit light, there are two approaches possible, other than the phosphorescent mechanism.

1. Delayed fluorescence by TTA (TTA, discussed in Section 2.4.2.4)
2. Upconversion of exciton from T_1 to S_1 state by thermal energy.

The second approach is called TADF (Thermally Activated Delayed Fluorescence). It has been known that several materials, such as porphyrin derivatives, show the TADF emission. However, as it is a thermally activated mechanism, which requires T_1 state to acquire energy ΔE_{ST} in Figure 2.33, the emission becomes smaller as the temperature is decreased. Therefore, to make TADF emission large enough to apply to the OLED device for realistic applications, it is necessary to secure the emission at lower temperature.

Adachi's group made a major breakthrough in this issue and made the TADF molecule usable for OLED device application for normal temperature range [49]. To minimize the temperature dependence of TADF method, it is necessary to minimize the ΔE_{ST}. The ΔE_{ST} can be written using exchange integral, K (see Section 2.3.3 for details) as follows.

$$\Delta E_{ST} = 2K \tag{2.71}$$

$$K = \iint \chi_A(1)\chi_B(2)\frac{1}{r_{12}}\chi_A(2)\chi_B(1)d\tau_1 d\tau_2 \tag{2.72}$$

This means that as the wavefunction overlap between ground state and excited state becomes smaller, ΔE_{ST} would be reduced. Therefore, orthogonal wavefunction would give smaller ΔE_{ST}, which will avoid the temperature dependence. However, there is an issue when such a molecule is to be designed. Radiation rate can be written by transition dipole moment μ as the following formula, which has very similar equation to (2.72); then the radiation would be decreased as ΔE_{ST} in most molecular designs.

$$I_{rad} \propto \mu^2 = \left\{ \iint \chi_A(1)\chi_B(2)r_{12}\chi_A(2)\chi_B(1)d\tau_1 d\tau_2 \right\}^2 \tag{2.73}$$

To make large enough radiation with minimum ΔE_{ST}, formula (2.72) must be small, while (2.73) is kept large. To make that happen, Adachi's group connected donor molecule and acceptor molecule by a chemical group with a steric hindrance. As a result, internal quantum efficiency near 100% was achieved using the TADF method [49].

Thus, TADF can make 100% internal quantum efficiency, but it is still ongoing R&D topic and many advanced approaches have been proposed. For example, TAF (TADF-assisted fluorescence) uses TADF molecule as an assist dopant of fluorescent OLED device. Molecule designs for small ΔE_{ST} generally tend to cause slower fluorescent emission, so the singlet state can be downconverted into triplet states before fluorescence. In the TAF system, fluorescent emitters with fast decay receive energy from TADF molecule and quickly decay, so it can avoid the high energy excitons to degrade the OLED device [50].

2.6.2 Other Types of Excited States

2.6.2.1 Excimer and Exciplex
A transient aggregate of excited singlet state ($^1C^*$) and ground state of the same molecule is called excimer. (It should be noted that excimer is an aggregate of excited state and is not applied to an aggregate of ground state.)

$$C + h\nu \rightarrow {}^1C * \xrightarrow{C} {}^1(C - C) *$$

Excimer is often observed in condensed aromatic compounds.

If similar situation happens between different molecules, it is called exciplex. The excited state of charge-transfer complex (discussed in Section 2.6.2.2) is a kind of this excimer. A high-efficiency emission mechanism using the exciplex formation has been proposed [51].

2.6.2.2 Charge-Transfer Complex
When an electron-donating compound and electron-accepting compound are located adjacently, new absorption peak appears due to the charge transfer between two molecules. This type of molecular compound is called charge-transfer complex or CT complex Figure 2.34 shows the energy diagram of CT complex.

CT complex can be used to enhance the charge injection from the electrode (Figure 3.4 in Section 3.1.1.1) and also can be used as charge generation layer to form a tandem device (discussed in Section 2.6.3).

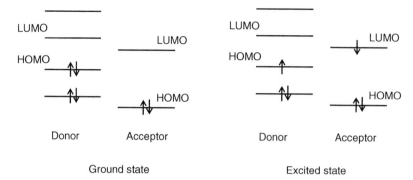

Figure 2.34 Ground state and excited state of CT complex.

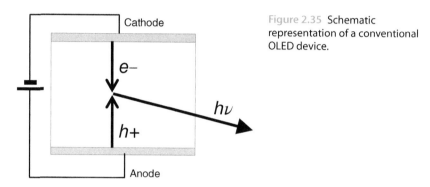

Figure 2.35 Schematic representation of a conventional OLED device.

With CT complex configuration, organic materials can have electrical conductivity (discussed in Section 8.1.5.1). Based on this mechanism, famous organic conductor TTF-TCNQ was developed, which shows metallic conduction mechanism (see Figure 8.8).

2.6.3 Charge Generation Layer

As discussed in Section 6.4.3, voltage drop is a significant issue in larger OLED displays. To reduce the voltage drop, tandem OLED device (also called multiphoton device) is effective, which increases the current efficiency (further discussed in Section 6.4.3).

Figure 2.35 shows a conventional OLED device, which sandwiches organic layers between the cathode and anode. Photons are generated in the organic layers by electron–hole recombination.

Figure 2.36a shows a two-unit tandem OLED structure, in which two organic layers are connected by an intermediate layer (known as a *connector layer* or a *charge generation layer* [CGL]) and the serial connection of organic layer units are sandwiched between the anode and cathode. Figure 2.36b shows a five-unit case.

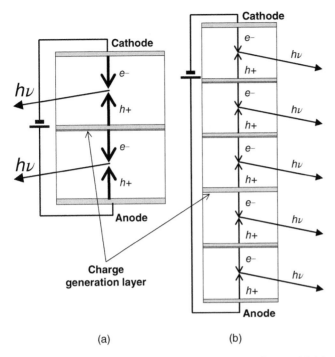

(a)   (b)

Figure 2.36 Schematic examples of a tandem OLED device with (a) two-layer and (b) five-layer serial connections.

With the tandem structure, by means of electrons and holes generated from the connector layer, photons are created in each organic layer unit, as if a charge were generating multiple photons. Ideally, the electric current of a tandem structure is the same as the current of the one-unit case and the voltage is N times greater than in the one-unit case. Therefore, current efficiency is N times greater, but power efficacy (luminous efficacy) is almost the same as that of the one-unit OLED device. (In power efficacy case, voltage increase cancels the emission increase in the equation.)

Ideally, with the same electric current, the N-unit tandem OLED device can deliver N times more luminance. The tandem device can be used in another way as well. The N-unit tandem OLED device can achieve the same luminance as the one-unit OLED device with $1/N$ times the electric current.

To make this operation happen, the CGL needs to have capability to produce electrons and holes and inject them into adjacent emissive units. Such CGLs are normally by either (1) conducting materials, such as ITO, or (2) insulating materials, such as V_2O_5 and 4 F-TCNQ, which forms CT complex with the adjacent material, such as aryl amine compound (discussed in Section 3.1.1.1) [52]. As both electrons and holes need to be injected from a common material to the adjacent layers, there occurs a large potential barrier for the carrier injection, especially for electrons. To facilitate the electron injection, alkhali dopants, organic dopants, and materials that induce vacuum-level shift are used.

References

1 A. Bemanose, M. Comte, and P. Vouaux, *J. Chim. Phys.* **50**:64 (1953); A. Bemanose, *J. Chim. Phys.* **52**:396 (1955); A. Bemanose and P. Vouaux, *J. Chim. Phys.* **52**:509 (1955).

2 H. Kallmann and M. Pope, *J. Chem. Phys.* **32**:300 (1960); H. Kallmann and M. Pope, *Nature* **186**(4718): 31 (1960).

3 M. Pope, H. Kallmann, and P. Magnante, *J. Chem. Phys.* **38**(8):2042 (1963).

4 M. Sano, M. Pope, and H. Kallmann, *J. Chem. Phys.* **43**(8):2920 (1965).

5 C. W. Tang and S. A. Van Slyke, Organic electroluminescent diodes, *Appl. Phys. Lett.* **51**(12):913–915 (1987).

6 T. Matsuura, *Photochemistry*, Kagaku Dojin, p. 11 (1970).

7 H.Inoue, et al., *Photochemistry I*, Maruzen, p. 103 (1999).

8 E. L. Murphy and R. H. Good, Thermionic emission, field emission, and the transition region, *Phys. Rev.* **102**(6):1464–1473 (1956).

9 P. S. Davids, S. M. Kogan, I. D. Parker, and D. L. Smith, Charge injection in organic light-emitting diodes: tunneling into low mobility materials, *Appl. Phys. Lett.* **69**(15): 2270 (1996).

10 B. K. Crone, I. H. Campbell, P. S. Davids, and D. L. Smith, Charge injection and transportation in single-layer organic light-emitting diodes, *Appl. Phys. Lett.* **73**(21):3162 (1998).

11 M. E. Kiziroglou, X. Li, A. A. Zhukov, P. A. J. de Groot, and C. H. de Groot, Thermionic field emission at electrodeposited Ni-Si Schottky barriers, *Solid-State Electron.* **52**(7):1032–1038 (2008).

12 H. Ishii, K. Sugiyama, E. Ito, and K. Seki, Energy level alignment and interfacial electronic structures at organic/metal and organic/organic interfaces, *Adv. Mater.* **11**:605 (1999).

13 N. Hayashi, H. Ishii, E. Itoh, and K. Seki, Electronic structure of organic/metallic interface observed by Kelvin method—band bending and large surface potential creation, *Appl. Phys. Mag.* **71**:1488 (2002).

14 H. Ishii, Metal-molecular interface electronic structure for organic molecular electronics, *Solid Phys.* **40**:375 (2005).

15 I. G. Hill, A. Rajagopal, A. Kahn, and Y. Hu, Molecular level alignment at organic semiconductor–metal interfaces, *Appl. Phys. Lett.* **73**:662 (1998).

16 I. G. Hill and A. Kahn, Energy level alignment at interfaces of organic semiconductor heterostructures, *J. Appl. Phys.* **84**(10): 5583 (1998).

17 C. Adachi et al., *Device Physics of Organic Semiconductors*, Kodansha Ltd., p. 33 (2012).

18 P. Mark and W. Helfrich, Space-charge-limited currents in organic crystals, *J. Appl. Phys.* **33**:205 (1962).

19 N. F. Mott and R. W. Gurney, *Electronic Processes in Ionic Crystals*, Oxford University Press, London and New York, 1940.

20 G. Horowitz, Organic field-effect transistors, *Advanced Materials*, **10**(5):365 (1998).

21 P. G. Le Comber and W. E. Spear, Electronic transport in amorphous silicon films, *Phys. Rev. Lett.* **25**:509–511 (1970).

22 C. W. Tang, S. A. Van Slyke, and C. H. Chen, Electroluminescence of doped organic thin films, *J. Appl. Phys.* **65**:3610–3616 (1989).

23 C. W. Tang, C. H. Chen, and R. Goswami, Electroluminescent Device with Modified Thin Film Luminescent Zone, US Patent 4,769,292 (1988).

24 M. A. Baldo, D. F. O'Brien, Y. You, A. Shoustikov, S. Sibley, M. E. Thompson, and S. R. Forrest, Highly efficient phosphorescent emission from organic electroluminescent devices, *Nature* **395**:151–154 (1998).

25 C. Ganzorig and M. Fujihira, A possible mechanism for enhanced electrofluorescence emission through triplet–triplet annihilation in organic electroluminescent devices, *Appl. Phys. Lett.* (**17**):3137 (2002).

26 Z. D. Popovic and H. Aziz, Delayed electroluminescence in small-molecule-based organic light-emitting diodes: evidence for triplet-triplet annihilation and recombination-center-mediated light-generation mechanism, *J. Appl. Phys.* **98**:13510 (2005).

27 S. Reineke, K. Walzer, and K. Leo, Triplet-exciton quenching in organic phosphorescent light-emitting diodes with Ir-based emitters, *Phys. Rev.* **75**:125328 (2007).

28 The IUPAC Gold book, http://goldbook.iupac.org/T06505.html.

29 S. M. Bachilo and R. B. Weisman, Determination of triplet quantum yields from triplet-triplet annihilation fluorescence, *J. Phys. Chem. A* **104**:7711-7714 (2000).

30 D. Kondakov, Characterization of triplet-triplet annihilation in organic light-emitting diodes based on anthracene derivatives, *J. Appl. Phys.* **102**:114504 (2007).

31 M. Funahashi, H. Yamamoto, N. Yabunouchi, K. Fukuoka, H. Kuma, and C. Hosokawa, Highly efficient fluorescent deep blue dopant for "super top emission" device, *SID Symposium Digest*, Vol.**39**, Issue 1, pp. 709-711, (2008).

32 X. Zhou, J. Blochwitz, M. Pfeiffer, A. Nollau, T. Fritz, K. Leo, Enhanced hole injection into amorphous hole-transport layers of organic light-emitting diodes using controlled p-type doping, *Adv. Funct. Mater.* **11**(4):10-314 (2001).

33 J. Kalinowski, W. Stampor, J. Mezyk, M. Cocchi, D. Virgili, V. Fattori, and P. Di Marco, Quenching effects in organic electrophosphorescence, *Phys. Rev. B* **66**:235321 (2002).

34 M. Baldo, C. Adachi, and S. Forrest, Transient analysis of organic electrophosphorescence. II. Transient analysis of triplet-triplet annihilation, *Phys. Rev. B* **62**:10967 (2000).

35 R. Meerheim, M. Furno, S. Hofmann, B. Lüssem, and K. Leo, Quantification of energy loss mechanisms in organic light-emitting diodes, *Appl. Phys. Lett.* **97**:253305 (2010).

36 A. Mikami, S. Nakajima, and A. Okada, Flexible polymer electroluminescent device with PVCz light-emitting layer on high-index of refraction substrate, *IDW 2002 Digest*, pp. 1139–1142 (2002).

37 D. Yokoyama, Molecular orientation in small-molecule organic light-emitting diodes, *J. Mater. Chem.* **21**:19187 (2011).

38 D. Yokoyama, Y. Setoguchi, A. Sakaguchi, M. Suzuki, and C. Adachi, Orientation control of linear-shaped molecules in vacuum-deposited organic amorphous films and its effect on carrier mobilities, *Adv. Mater.* **20**(3):386-391 (2010).

39 M. Era, T. Tsutsui, and S. Saito, Polarized electroluminescence from oriented p-sexiphenyl vacuum-deposited film, *Appl. Phys. Lett* **67** (17):2436-2438 (1995).

40 T. Cremoux, M. Dussauze, E. Fargin, T. Cardinal, D. Talaga, F. Adamietz, and V. Rodriguez, Trapped molecular and ionic species in poled borosilicate glasses: toward a rationalized description of thermal poling in glasses, *J. Phys. Chem. C* **118**:3716–3723 (2014).

41 P. Dyreklev, M. Berggrem, W. Inganas, M. Andersson, O. Wennerstrom, and T. Hyertberg, Polarized electroluminescence from an oriented substituted polythiophene in a light emitting diode, *Adv. Mater.* **7**:43–45 (1995).

42 T. Tsujimura, K. Furukawa, H. Ii, H. Kashiwagi, M. Miyoshi, S. Mano, H. Araki, and A. Ezaki, World's first all phosphorescence OLED product for lighting application, *IDW 2011 Digest* (2011).

43 T. Yoshioka, K. Sugimoto, K. Katagi, Y. Kitago, M. Tajima, S. Miyaguchi, T. Tsutsui, R. Iwasaki, and Y. Furukawa, An improved method for lifetime prediction based on decoupling of the joule self-heating effect from coulombic degradation in accelerated aging tests of OLEDs, *SID 2014 Digest*, pp. 642-645 (2014).

44 H. Aziz and Z. Popovic, Degradation phenomena in small-molecule organic light-emitting devices, *Chem. Mater.* **16**(23):4522-4532 (2004).

45 P. Burrows, B. Bulovic, S. Forrest, L. Sapochak, D. McCarthy, and M. Thompson, Reliability and degradation of organic light emitting devices, *Appl. Phys. Lett.* **65**(23):2922-2924 (1994).

46 D. Kondakov and R. Young, Role of chemical transformations of hole transport materials in operational degradation of the current generation of highly efficient fluorescent OLEDs, *SID Symposium Digest*, Vol. **40**, Issue 1, pp. 687-690 (2009).

47 R. Meerheim, S. Scholz, S. Olthof, G. Schwartz, S. Reineke, K. Walzer, and K. Leo, Influence of charge balance and exciton distribution on efficiency and lifetime of phosphorescent organic light-emitting devices, *J. Appl. Phys.* **14**:014510 (2008).

48 F. So and D. Kondakov, Degradation mechanisms in small-molecule and polymer organic light-emitting diodes, *Adv. Mater.* **22**:3763-3777 (2010).

49 H. Uoyama, K. Goushi, K. Shizu, H. Nomura, and C. Adachi, Highly efficient organic light-emitting diodes from delayed fluorescence *Nature* **492**:234–238 (2012).

50 H. Nakanotani, T. Higuchi, T. Furukawa, K. Masui, K. Morimoto, et al., High-efficiency organic light-emitting diodes with fluorescent emitters, *Nat. Commun.* **5**:4016 (2014).

51 S. Seo, T. Takahashi, H. Nowatari, S. Hosoumi, T. Ishisone, T. Watabe, S. Mitsumori, N. Ohsawa and S. Yamazaki, Efficiency enhancement in phosphorescent and fluorescent OLED utilizing energy transfer from exciplex to emitter, *SID 2015 Digest*, Vol. **46**, Issue 1, pp. 605-608 (2015).

52 J. Kido, High performance OLEDs for displays and general lighting, *SID Symposium Digest of Technical Papers*, Vol. **39**, Issue 1, pp. 931–932 (2008).

3

OLED Manufacturing Process

The process for manufacturing high-quality OLED devices is discussed in this chapter.

3.1 MATERIAL PREPARATION

3.1.1 Basic Material Properties

As discussed in Section 2.3, OLED devices consist of multiple layers that have different functions. It is very important to design layers by selecting material in terms of their inherent properties and compatibility (e.g., with respect to their HOMOs and LUMOs) with other layers.

3.1.1.1 Hole Injection Material

The term *hole injection* refers to the injection of holes from an anode, namely, ejection of electrons to the anode. The mechanism is discussed in Section 2.3.6. Hole injection material is used to reduce the height of the barrier (also termed a *potential barrier*) against hole injection from anodes [1, 2]. The material layer involved in this process is called the *hole injection layer* (HIL).

Hole injection material is often selected so that the HOMO level is located between the HOMO of the hole transportation material (HTM) and the work-function of the anode (Figure 3.1).

Figure 3.2 shows typical hole injection materials.

There is another type of HIL used popularly, such as $HAT(CN)_6$ (Figure 3.3). $HAT(CN)_6$ is known to form charge-transfer complex (discussed in Section 2.6.2.2.) with aryl amine compounds, such as NPB. Figure 3.4 shows the case when $HAT(CN)_6$ is used as HIL and NPB is used as HTL. Electrons are extracted from the interface due to charge-transfer mechanism and conduct toward the anode. Also holes, created by electron extraction, conduct toward the cathode . (Act as a kind of charge generation layer, discussed in Section 2.6.3.) As a result, effective hole injection can make happen for high-performance OLED operation.

OLED Display Fundamentals and Applications, Second Edition. Takatoshi Tsujimura.
© 2017 John Wiley & Sons, Inc. Published 2017 by John Wiley & Sons, Inc.

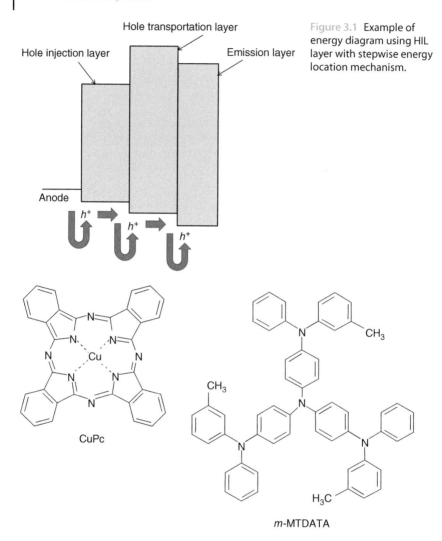

Figure 3.1 Example of energy diagram using HIL layer with stepwise energy location mechanism.

Figure 3.2 Example structures of hole injection material.

3.1.1.2 Hole Transportation Material

HTM is used to transport holes to the emission layer (EML) [3]. This layer is called the *hole transport layer*, which is usually designed with a higher exciton energy than EML to ensure that the excitation energy of the EML is not transferred to the transport layer.

Figure 3.5 shows the structures of typical HTMs.

3.1.1.3 Emission Layer Material

The layer from which light is emitted is referred to as the EML [4, 5]. The electron transportation and hole transport layers can also function as EMLs.

Figure 3.3 Structure of HAT(CN)$_6$. hole injection material.

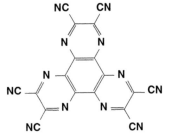

Figure 3.4 Example of energy diagram using HIL layer with charge-transfer complex mechanism.

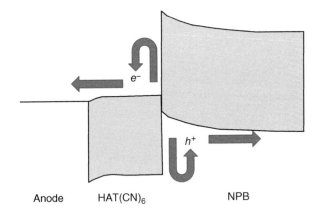

Anode   HAT(CN)$_6$ NPB

Host material normally has light emission capability, but doping (Section 2.4.2.3) of the host material is a very useful approach when emission from the host material causes concentration quenching (quantum yield loss at high concentration) or when the color needs fine-tuning. In such cases, the bandgap of the host material should be wider than that of the dopant.

Figure 3.6 shows typical fluorescent light-emitting materials.

Phosphorescent materials [6, 7] are also used for the EML. Normally, light emission from the triplet state is prohibited. However, with the use of heavy-metal complexes containing iridium and platinum, for example, light emission is allowed as a result of spin–orbit coupling, due to the heavy-metal effect (discussed in Section 2.4.2.3). Figure 3.7 shows the structure of a typical phosphorescent dopant. For phosphorescent emission, the triplet bandgap of the host needs to be wider than that of the dopant, so a host material specifically designed for phosphorescent emission is sometimes used (Figure 3.8) [8].

3.1.1.4 Electron Transportation Material and Charge Blocking Material

Electron transportation material is used to transport electrons injected from the cathode [9–11]. This layer, called the *electron transport(ation) layer* (ETL), is sometimes used with the *electron injection layer* (EIL), such as LiF, which induces vacuum-level shift (Section 2.3.6.3) when it is allocated with organic

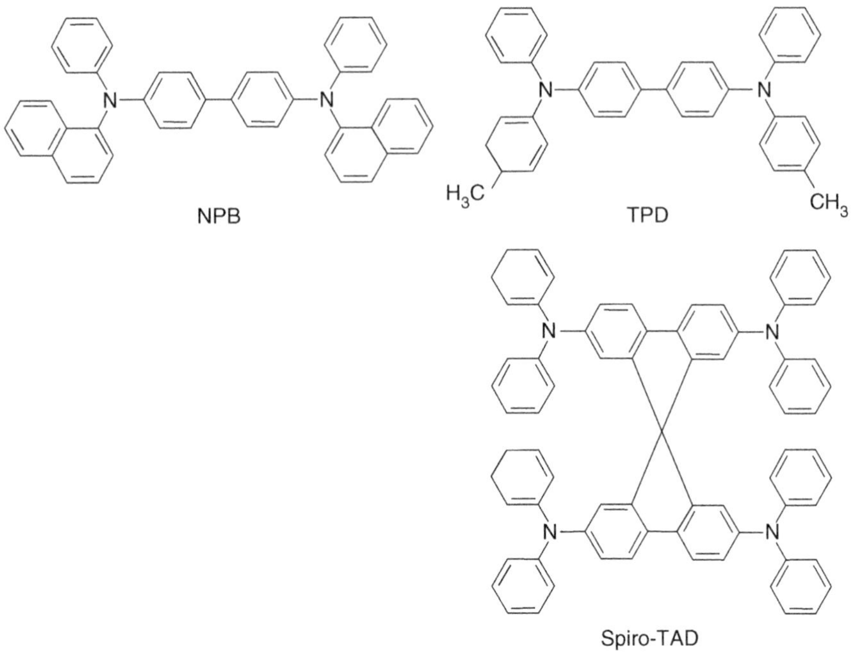

NPB

TPD

Spiro-TAD

Figure 3.5 Example structures of hole transportation material.

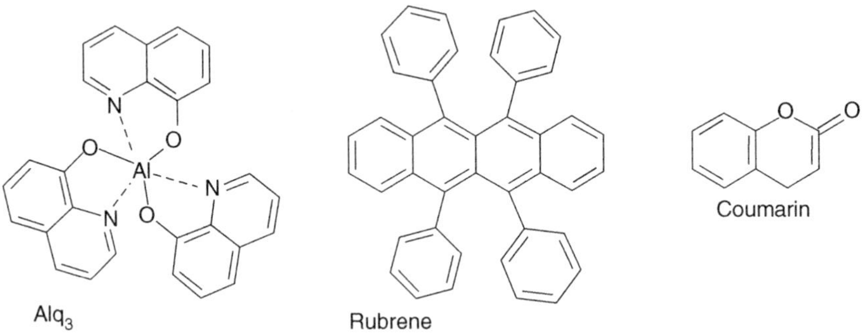

Alq₃

Rubrene

Coumarin

Figure 3.6 Example structures of fluorescent light-emitting materials.

materials, or alkhali dopants, such as Li, Ca, or organic materials such as Liq₃ with smaller electron affinity than the electrode, to increase the electron injection capability (the injection mechanism is discussed in Section 2.3.6.). Sometimes, the ETL is used without EIL when the cathode workfunction matches with the LUMO of the ETL.

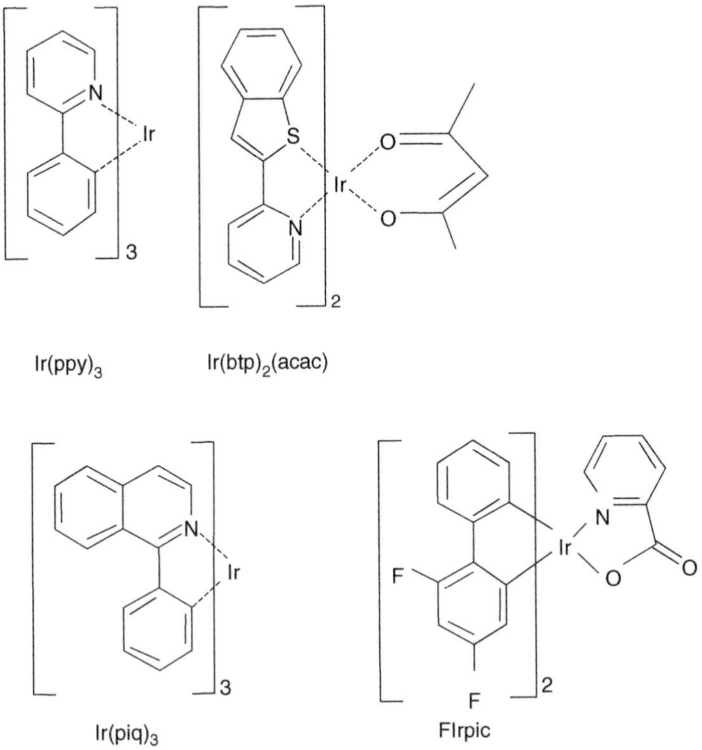

Ir(ppy)₃   Ir(btp)₂(acac)

Ir(piq)₃   FIrpic

Figure 3.7 Example structures of phosphorescent dopant materials.

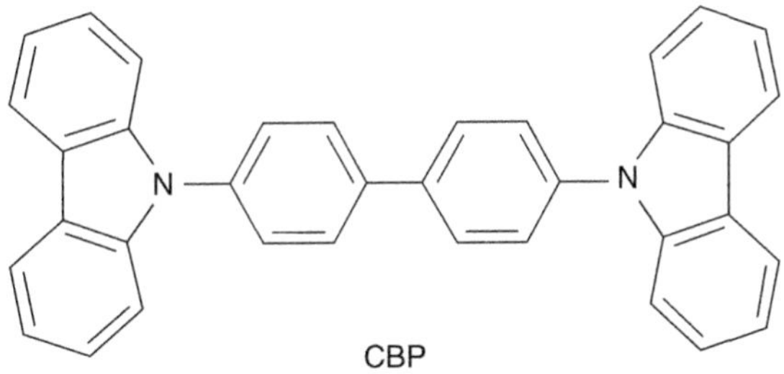

Figure 3.8 Example structure of a phosphorescent host material.

CBP

Just like the hole transport layer, the electron transport layer should have a higher exciton energy than that of the EML. The electron transport layer also sometimes blocks exciton and hole movement to enhance emission efficiency.

Figure 3.9 shows a typical example of the electron transport layer. BCP(bathocuproine) is also known as a popular *hole blocking material* (HBM). The hole blocking layer (HBL), a layer of HBM material, is used to block the hole transfer so that the charge is accumulated in the EML layer.

Alq₃

Bathocuproine (BCP)

Figure 3.9 Example structures of electron transportation material.

Figure 3.10 shows a simple two-layered OLED device without HBL. Generally speaking, organic material conducts more holes than electrons. Therefore, the amount of holes accumulated in the hole transportation layer/EML interface is normally higher than the amount of electrons accumulated in the same interface. The electric field caused by the accumulated holes facilitates the electron transport, which results in good carrier balance. The holes that meet with electrons would successfully recombine and most of them would create photons for emission. However, especially when EML has high hole conduction, such as due to Ir-complex molecule doping, excessive holes increase the leakage current. Such leakage current reduces the OLED device efficacy and also causes undesired light emission from unexpected layers.

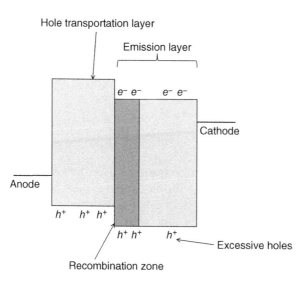

Figure 3.10 Schematic representation of energy diagram using two-layer OLED device without hole blocking layer.

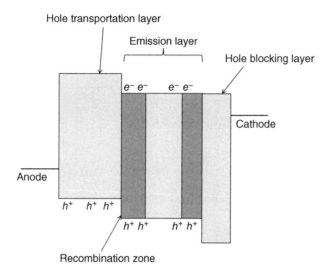

Figure 3.11 Schematic representation of energy diagram using two-layer OLED device with hole blocking layer.

To circumvent the situation, HBL is used. In the case of OLED device shown in Figure 3.11, excessive holes meet with energy barrier at EML/HBL interface. As the electric field created by blocked holes suppresses the hole injection from the hole transport layer to EML, the amount of holes in EML is maintained. As a result, two recombination zones are established. Due to the recombination location control, efficacy loss due to leakage current can be suppressed. Also emission from unexpected layers can be avoided. By accumulation of charge, effective electron–hole recombination can be made to achieve high efficiency by means of better carrier balance, as discussed in Section 2.4.1.

3.1.2 Purification Process

The characteristics and lifetime of an OLED device depend significantly on the purity of the materials used. For this reason, sublimation purification is often used to increase the purity of the materials. Figure 3.12 shows a simplified layout for typical equipment employed for sublimation purification. This equipment consists of a quartz tube, a metal tube attached to the quartz tube (the metal tube is used to set the temperature gradient), and a band heater.

Several grams of material to be purified are heated by the band heater. The greater the distance between the material and the band heater, the lower the temperature, so sublimed material is transferred to the lower-temperature location. Nonvolatile impurities remain in the boat, and high-mass material is extracted in the high-temperature region. Low-mass material is extracted in the low-temperature region, so mass separation is possible using this

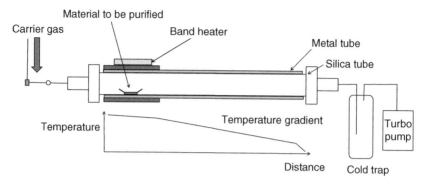

Figure 3.12 Schematic representation of sublimation purification equipment.

method. During OLED device fabrication, even if the material is uniform, any impurities that it contains can significantly influence the characteristics of the OLED device. In that sense, sublimation purification is very important.

3.2 EVAPORATION PROCESS

3.2.1 Principle

The *evaporation* process [12] involves conversion of a solid or a liquid to the vapor state by heating the material in an evaporation source. A material that is transformed directly into vapor, bypassing the liquid phase, is called a sublimation-type material. On the other hand, a material type that turns into liquid and then evaporates is called a melting-type material (further discussed in Section 3.2.2.1). The simplest evaporation source is called a "boat": a boat-shaped vessel made of metal, such as tungsten, which also acts as a resistive heating source to raise the temperature of the material in the vessel to vaporize it. The heat can also be applied by electron beam (EB) evaporation; however, a high-energy electron, ion, or X-ray can damage the TFT substrate and OLED device, so resistive heating is more commonly used. The basic structure of evaporation equipment, using a boat for the evaporation source, is shown in Figure 3.13. For organic materials, an evaporation crucible is popular for manufacturing (Figure 3.14), as it can apply heat more uniformly than a boat. In the case of the crucible, heat is applied by the resistive-heating filament, which is normally placed outside the crucible.

The basic parameters of the evaporation process can be described as follows. By applying the kinetic theory of gases, we can express the number of molecules evaporated per unit evaporation area per unit time ($/m^2s$), when the material

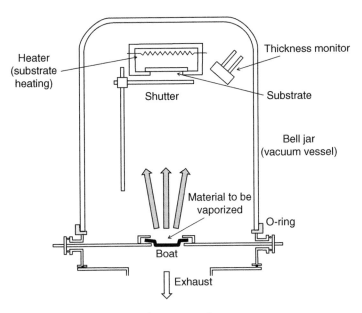

Figure 3.13 Basic structure of an evaporation source.

is heated and then extracted by evaporation, as

$$J = \frac{1}{4}\alpha n\bar{v}$$
$$= 2.6 \times 10^{24} P\sqrt{MT} \tag{3.1}$$

where n is the number of molecules per unit volume, $\bar{v}$ the average velocity, P the pressure, M the molecular mass, and α the cohesion coefficient (α is almost 1).

In Eq. (3.1), as the temperature increases, vapor pressure also increases, so the deposition rate is increased. An OLED device normally requires precise evaporation speed control, so a Knudsen cell (Figure 3.14) (sometimes called a "K cell," a crucible used to apply more uniform temperature gradient) is often used for manufacturing.

The mean free path can be expressed by the equation

$$\lambda = \frac{1}{\sqrt{2}\pi nD^2} \tag{3.2}$$

where λ is the mean free path (m), n the molecular number density in vacuum, and D the molecular diameter (m). For example, if the molecular diameter is 1.4 nm (Alq_3), the vacuum level must be larger than 10^{-4} Pa when the distance between crucible and substrate is 30 cm. The purpose of a high vacuum is to

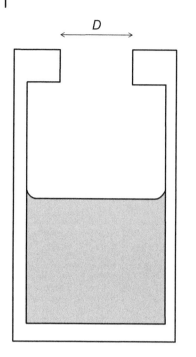

lengthen the mean free path to avoid any impurities being included in the film, such as oxygen, which causes oxidization.

The amount of material deposited on the substrate is determined by the evaporation angle and source–substrate distance. Here, we assume that the evaporated molecule originated from a point source. Isotropic evaporation is also assumed.

The number of molecules evaporated into the solid angle $d\omega$ is

$$dN = \frac{N d\omega}{4\pi} \tag{3.3}$$

where N is the total number of evaporated molecules per second (Figure 3.15).

The frequency of incident evaporation molecules at point A in Figure 3.16 is

$$n = \frac{N \cos \phi}{4\pi r^2} \tag{3.4}$$

Here, we assume that the substrate is located immediately above the evaporation source at a distance h.

In the θ direction, h and θ can be described as

$$h = r \cos \theta$$
$$\theta = \phi$$

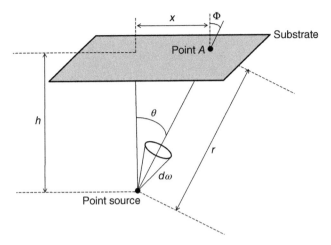

Figure 3.15 Incident evaporation molecule to substrate at point A from point source.

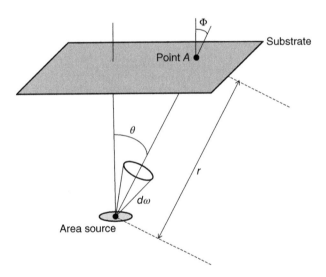

Figure 3.16 Incident evaporation molecule to substrate at point A from area source.

Using Eq. (3.4), we note that the thickness of the deposited layer is proportional to $1/\cos^3\theta$. The thickness distribution $d(x)$, at a distance x from the right (point A in Figure 3.16) below the evaporation source, can be expressed as

$$d(x) = d(0)\left\{1 + \frac{x^2}{h^2}\right\}^{-(3/2)} \tag{3.5}$$

On the other hand, if a very small area evaporation source, not a point source, is assumed, then, according to the rule of cosines, we obtain

$$dN = \frac{N \cos \theta \, d\omega}{\pi} \tag{3.6}$$

Therefore, we can express n as

$$n = \frac{N \cos \theta \cos \phi}{\pi r^2} \tag{3.7}$$

The thickness distribution is now proportional to $1/\cos^4\theta$. Then, according to this formula, the thickness distribution $d(x)$ of area source (Figure 3.16) at a distance x can be expressed as follows:

$$d(x) = d(0) \left\{ 1 + \frac{x^2}{h^2} \right\}^{-2} \tag{3.8}$$

The simulated thickness distribution for the point source (Eq. [3.5]) and for the area source with a very small area (Eq. [3.8]) is plotted in Figure 3.17. Normally, an OLED device requires precise thickness control with variation within several percent of thickness (requirements for microcavities are tighter), so the variation shown in Figure 3.17 renders the thickness distribution in OLED devices nonuniform and thus unsuitable for this display application.

To correct the variation in thickness distribution, the following strategies are recommended:

- Increasing source to substrate distance (T/S); this distance is also known as *target–source distance* (T/S).
- Rotation of substrate.
- Offset positioning from the rotation center (Figure 3.18).

As the T/S of a point source needs to be long, a large amount of material is deposited on the chamber wall and shield plate, so the material utilization level tends to be low.

3.2.2 Evaporation Sources

3.2.2.1 Resistive Heating Method

In resistive heating, the most frequently used evaporation method, Joule heating due to electric current flow through the resistance of a high-melting-point metal, such as tungsten, tantalum, or molybdenum, is used to cause evaporation of the material. There are two methods of implementation: (1) using high-melting-point metal as a filament to create heating and (2) using the metal as a boat itself. If method 2 is used, it is important to ensure compatibility between the boat material and the evaporation material in order to avoid interaction between these two materials.

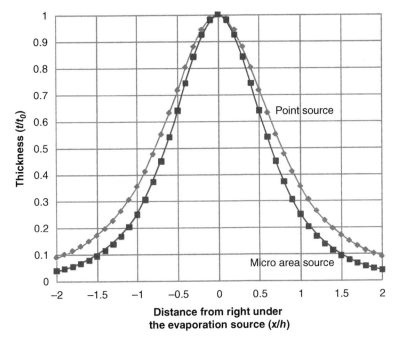

Figure 3.17 Thickness distribution in the case of motionless evaporation.

In another method, in the case of filament implementation, a boron nitride (BN), quartz, or graphite crucible is heated by a tungsten filament heater.

It is also possible to deposit a mixture such as by the doping method using multiple crucibles. (Normally, multiple materials contained in one crucible cannot be evaporated as a mixture. This is because the evaporation pressure values of different materials vary, so only one material will be evaporated while the others are not, as in a distillation process.)

All OLED materials can be classified into one of the following two groups: (1) melting-type material, which evaporates after melting, and (2) sublimation-type material, which sublimes without melting, as discussed in Section 3.2.1. The evaporation source should be selected according to the evaporation parameters of the specific material.

If evaporation is carried out by only one evaporation source for 1 week, the OLED device characteristics may change during operation of the evaporation equipment or the material to be evaporated may become depleted. To avoid such problems, a revolver (also known as a *turret*) mechanism (Figure 3.19) is often used.

Figure 3.20 shows the transformation of an evaporation source material during operation. OLED materials seldom have high thermal conductance, so the

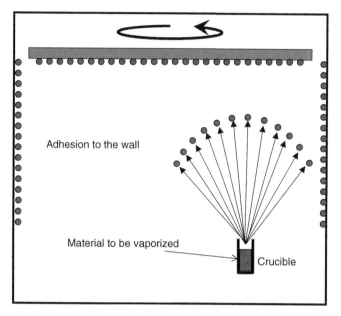

Adhesion to the wall

Material to be vaporized

Crucible

Figure 3.18 Schematic example of evaporation with offset and substrate rotation.

Figure 3.19 Evaporation equipment using a revolver apparatus.

Figure 3.20 Schematic representation of material in an evaporation source.

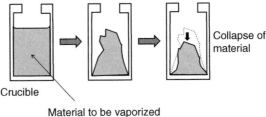

Crucible

Material to be vaporized

Collapse of material

Mixture of material and thermoball

Crucible

material near the wall tends to be evaporated more rapidly than that in the center area. Moreover, the stacked, accumulated material may suddenly collapse, resulting in a significant change in evaporation rate, which is problematic for manufacturing.

As the material near the wall is consumed more rapidly, the heater temperature must be higher than the normal evaporation temperature of the material to ensure adequate evaporation speed. However, this higher temperature can degrade the material's properties.

To overcome this problem, a material known as a *thermoball* (see the lower diagram in Figure 3.20; photo in Figure 3.21), which has a high thermal conductance, with thermal stability at the evaporation temperature, is sometimes mixed with the material to be vaporized. Because of the high thermal conductance, issues such as sudden change in evaporation speed or material decomposition due to overheating can be avoided.

A linear evaporation source (see Figures 3.22 and 3.23) is also used to increase manufacturability [13, 14]. With a linear source, uniform deposition can be achieved even when the source is very close to the substrate, so this feature is advantageous in terms of material utilization. However, as the source is close to the substrate, this can lead to OLED device degradation due to radiative heating and shadow mask deformation as a result of thermal expansion. Also, because of the thermal conductance problems of the material, it is not easy to evaporate materials uniformly along the whole length of the linear source, so material utilization is not always as high as expected.

3.2.2.2 Electron Beam Evaporation

Use of the resistance heating method can result in contamination of the evaporated material. In EB evaporation, an accelerated EB is emitted toward the target material to evaporate it. This method can be used to produce high-quality

Figure 3.21 A thermoball.

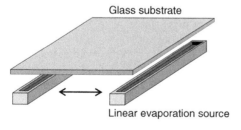

Glass substrate

Figure 3.22 Thin-film evaporation by linear source.

Linear evaporation source

films but may also result in damage to the TFT and OLED device due to the ultraviolet light and X-ray radiation created.

3.2.2.3 Monitoring Thickness Using a Quartz Oscillator
In an actual full-length evaporation operation, the evaporation rate fluctuates. To ensure accurate thickness measurement, feedback control is necessary for

Figure 3.23 Linear source evaporation equipment.

monitoring the thickness. A quartz oscillator is often used to measure the thickness (Figure 3.24).

In thickness measurement using a quartz oscillator, the oscillation frequency of the quartz is reduced because of the effective change in mass due to the film formation. As the Q value of the quartz oscillator exceeds 10^6, this method has about 0.5 nm sensitivity.

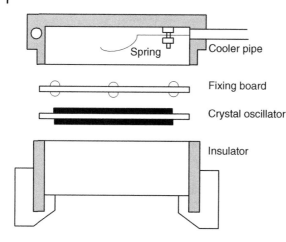

Figure 3.24 Schematic representation of a thickness monitoring apparatus using a quartz crystal oscillator.

The correlation between resonant frequency f_q of the quartz oscillator and thickness t_q can be expressed as follows:

$$f_q \cdot t_q = \frac{v_q}{2} = N_q = 1.69 \times 10^3 \quad (m/s) \tag{3.9}$$

If the oscillation frequency of the quartz oscillator without film formation is f_c, then

$$\rho_f t_f = \rho_q t_q \frac{f_q - f_c}{f_q} \tag{3.10}$$

where ρ_q and ρ_f are density of the quartz oscillator and density of the deposited film, respectively, and t_q and t_f are thicknesses of the quartz oscillator and the deposited film.

When the film is thick, this equation is no longer valid, and

$$\rho_f t_f = \rho_q t_q \frac{f_q - f_c}{f_c} \tag{3.11}$$

should be used instead.

Here, the Z ratio is defined as follows:

$$Z = \frac{\rho_q v_q}{\rho_f v_f} \tag{3.12}$$

Then

$$\rho_f t_f = \frac{\rho_q t_q}{\pi Z(1 - F)} \tan^{-1}\{Z \tan(\pi F)\} \tag{3.13}$$

In this equation, we have

$$F = \frac{f_q - f_c}{f_q} \tag{3.14}$$

According to this equation, thickness can be calculated from the frequency difference. Although the crystal thickness monitoring method is able to provide continuous measurement during operation, to keep the measurement accuracy, frequent quartz replacement is required, which causes suspension of the manufacturing process. Therefore, newer deposition method is proposed, which uses ellipsometry or a Pirani gauge (Section 5.3.1) to measure the thickness during operation, circumventing this issue.

3.3 ENCAPSULATION

3.3.1 Dark Spot and Edge Growth Defects

To produce high-quality OLED displays, defects should be avoided in terms of manufacturing and reliability. An OLED is highly vulnerable to water damage, so it is important to protect the device from humidity.

Figure 3.25 shows a defect called a *dark spot (also called "black spot" or "black dot")*, which is believed to be caused by the cathode peeling or by organic material degradation due to the presence of pinholes (Figure 3.26) created by several mechanisms, such as irregularities of the substrate surface or of the anode, or the protrusion structure of organic film [15, 16].

When encapsulation is insufficient, the nonemissive area gradually increases, as shown in Figure 3.27. As a result, effective pixel size decreases; this is called *pixel shrinkage* or *edge growth.*

Moisture in the OLED display panel does not only come from the outside. Especially when polymer resin is used for planarization or edge insulation, the material and baking conditions should be carefully investigated to ensure that the presence of the polymer does not result in excess moisture.

Schaer et al. have made an experiment to investigate how dark spots are grown in controlled atmosphere [17]. In the experiment, it was found that the degradation of OLED in water vapor is very much different from the degradation in pure oxygen, and the degradation speed by water is faster by three orders of magnitude than oxygen. Normally, aluminum cathode shows excellent protection against water and oxygen; however, when the pinholes exist due to particles, dark spots are created. In the case of degradation due to water vapor, delamination of cathode is observed, which is possibly due to the bubble burst, caused by hydrogen evolution, which is created by electrochemical reduction of water. The delamination of cathode inhibits the electron injection of cathode to organic layers, which prevents the OLED emission in the region.

The growth of dark spot follows the following equation:

$$D(t) = D_0 e^{kt}$$

where D is the dark spot diameter, D_0 the initial dark spot diameter, and k the growth rate.

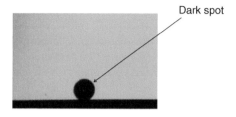

Dark spot

Figure 3.25 Dark spot formation.

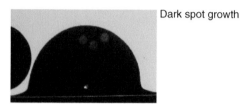

Dark spot growth

Cathode material

Particle

Organic material

Anode

Glass

Figure 3.26 Schematic representation of dark spot formation mechanism.

According to [17], the growth rate of dark spot is

$$k = 2 \times 10^{-4} \ [s^{-1}]$$

This value changes depending on the condition.

As the time goes on, visible dark spot number is increased and the appearance of a display is deteriorated, so it is very important to avoid internal and external moisture in manufacturing high-quality OLED displays.

3.3.2 Light Emission from the Bottom and Top of the OLED Device

An encapsulation strategy to protect an OLED display panel from moisture and oxygen is very important for fabricating a good OLED display.

Figure 3.28 shows light emission structures using both bottom and top emission methods.

Figure 3.27 Edge growth (a) before and (b) after high-temperature/high-humidity testing (sample with encapsulation).

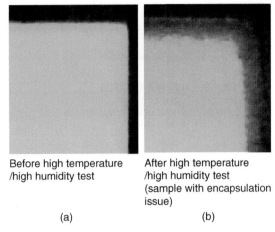

Before high temperature /high humidity test

After high temperature /high humidity test (sample with encapsulation issue)

(a)   (b)

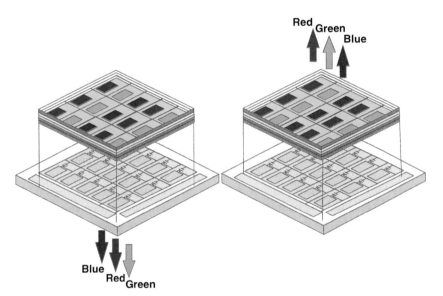

Figure 3.28 Light emission from (a) bottom and (b) top of the OLED structure.

3.3.3 Bottom Emission and perimeter sealing

Bottom emission structure has been used since the first commercialization of OLED display and is a mature technology.

For the bottom structure, opaque aluminum is used for the upper electrodes (normally cathodes) of the OLED device, and a transparent electrode such as ITO is used for the lower electrodes (normally anodes). In that case, the light

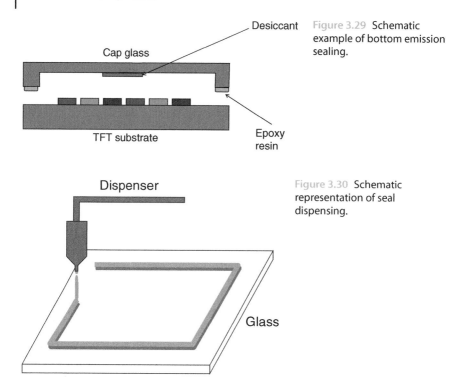

Figure 3.29 Schematic example of bottom emission sealing.

Figure 3.30 Schematic representation of seal dispensing.

generated in the OLED device passes through the glass and is emitted to out-side. In the bottom emission structure, perimeter sealing plus a desiccant have been used for most of the products. A desiccant is inserted between the OLED device substrate and the counter substrate using a hollow device called a *cap glass* or a cap metal, to absorb any moisture in the display device (Figure 3.29). Epoxy resin is distributed by a dispenser (Figure 3.30) or by screen printing along the edge of the cap substrate to form the perimeter seal, which is then attached to the OLED device substrate. The perimeter seal is cured by UV or heat (Figure 3.31).

3.3.4 Top Emission

To allow a high-resolution display to have enough aperture ratio in a pixel, top emission structure is often used. Especially, when an active-matrix OLED display needs to have compensation circuit (discussed in Section 6.4.4.), the cir-cuitry occupies significant portion of a pixel, so the aperture ratio becomes very small in the case of bottom emission. Using top emission, the pixel electrode can be placed above the circuitry, so the aperture ratio is not reduced regardless of the complexity of the circuitry.

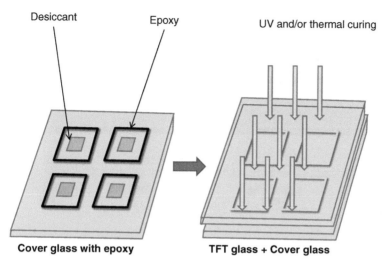

Figure 3.31 An example of perimeter-sealing process (bird's-eye view).

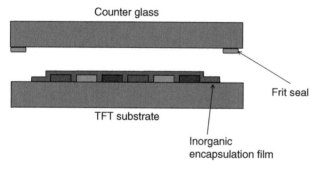

Figure 3.32 Schematic example of top emission encapsulation.

In a top emission structure, it is not possible to place opaque desiccants within the emissive device. (Use of a transparent desiccant has been proposed but is not widely applied in actual manufacturing.) Without a desiccant, more strict moisture control is necessary than for a bottom emission device. Thin-film encapsulation is used for many of the top emission OLED devices in combination with perimeter sealing, face sealing, and frit sealing technologies.

Figure 3.32 shows an example of top emission structure encapsulation, which uses a combination of thin-film encapsulation and frit sealing.

3.3.5 Encapsulation Technologies and Measurement

To encapsulate OLED device from moisture and oxygen, several techniques are used. Also WVTR measurement is used to evaluate the barrier performance.

Inorganic film (SiN$_x$, SiO$_x$N$_y$ etc.)

Organic film

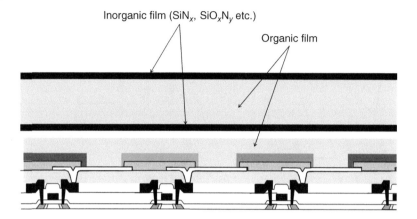

Figure 3.33 Schematic representation of film encapsulation by inorganic–organic repetition.

3.3.5.1 Thin-Film Encapsulation

Inorganic films, such as SiN$_x$ or SiN$_x$O$_y$, are known to protect OLED devices from moisture and oxygen. (In the case of triplet emission, exciton is susceptible to quenching by oxygen's triplet state, as discussed in Section 2.4.2.) However, the inorganic films for barrier purpose usually have pinholes, which leak the moisture and oxygen. To circumvent the situation, alternate deposition of inorganic and organic films (Figure 3.33) is often used.

Organic films tend to have good step coverage but poor protection capability against humidity. On the other hand, inorganic film tends to have poor step coverage but shows great barrier property when it once covers. As such, organic/inorganic film repetition can achieve good barrier performance despite the existence of pinholes [18].

Prins et al. made a modeling of humidity penetration through barrier film pinholes [19]. With organic/inorganic repetition structure, organic layer does not block the humidity/oxygen penetration. Inorganic layer basically blocks the humidity/oxygen penetration, but pinholes in inorganic layer leak the moisture/oxygen. As pinholes of different inorganic layers exist in different locations, the humidity/oxygen needs to travel long distance until it gets out of the barrier film structure. Due to the long distance, the conductance of humidity/oxygen flux is suppressed and it enhances the barrier property. This is called "Labyrinth effect" or "Tortuous path effect."

The technique can also be applied to barrier film (discussed in Section 8.1.4), which is normally fabricated by depositing barrier layers on plastic film.

In most cases, the thin-film barrier layers are normally deposited by CVD (Figure 6.7), sputtering (Figure 6.7) or ion plating. Figures 3.34 and 3.35 show the SiO$_2$ film deposited on aluminum wiring with and without taper shape,

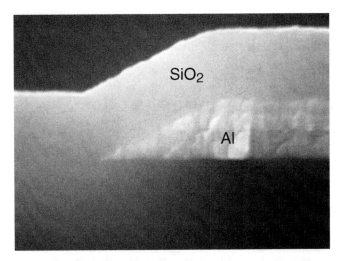

Figure 3.34 SiO$_2$ film deposited by CVD on tapered aluminum [20].

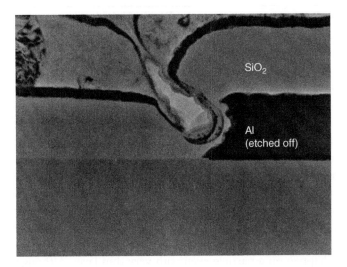

Figure 3.35 SiO$_2$ film deposited by CVD on nontapered aluminum [20].

respectively [20]. It can be clearly shown that the coverage of CVD SiO$_2$ film is not good enough when wiring is not tapered (Figure 3.35). Same situation can happen when film is deposited on a particle. If coverage is good enough, the particle can be covered so that the number of leakage path can be minimized. However, if the coverage is insufficient, water leakage path is created, which causes dark spot growth. It is known that normally CVD film coverage is better

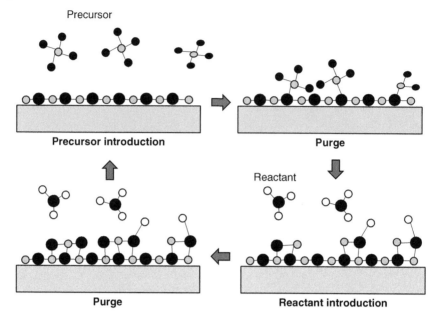

Figure 3.36 Schematic depiction of the reaction mechanism of the atomic layer deposition (ALD) method.

than that of film by sputtering method. Said that, even CVD does not provide perfect coverage over particles.

To improve the coverage issue on a particle, ALD (atomic layer deposition) deposition is attracting attention.

Figure 3.36 illustrates the reaction mechanism of the ALD method. Following introduction of the precursor gas to the film, the monolayer molecule is adsorbed. After the precursor gas is purged by the pump, reactant gas is introduced that reacts with reactant adsorbed precursor to form a monolayer film. This two-step procedure is repeated to form a high-quality film. As the interaction happens even beneath the particle, ALD deposition can make very good coverage happen on a particle even with very thin layer. Although this method yields high-quality film, the deposition rate obtained using the ALD method is low. To circumvent this issue, parallel processing of the ALD method, such as batch processing, can be used. To further improve the productivity, a high-speed ALD method termed *spatial ALD* (Figure 3.37) is attracting attention as a possible means of addressing this issue [21]. Spatial ALD uses a principle to transfer substrate on various gas flows so that the monolayer film is deposited sequentially. Spatial isolation of reactive gas enables layer-by-layer deposition without a time-consuming purging process,

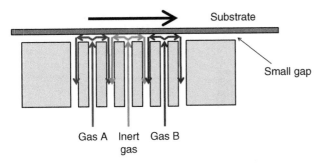

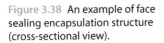

Figure 3.37 Schematic depiction of spatial ALD system.

Figure 3.38 An example of face sealing encapsulation structure (cross-sectional view).

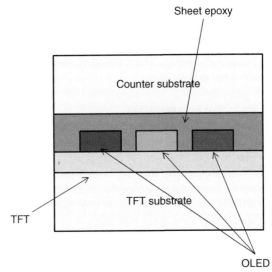

which is necessary in the case of the normal ALD method. Therefore, spatial ALD offers both high productivity and high-quality film formation capability.

3.3.5.2 Face Sealing Encapsulation

The device in Figure 3.38 has counterplate and substrate sandwiching the OLED device and is glued together by the blanket sealing material. The structure is called "face sealing [22]." The face sealing does not have any cavity in its structure, so it very well matches with flexible OLED device. Also, as the display size becomes larger, it becomes difficult to sustain the counterplate by peripheral structure. In such cases, face sealing technology can bring mechanically robust feature.

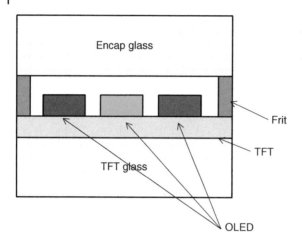

Figure 3.39 An example of face-frit encapsulation structure (cross-sectional view).

Sheet adhesive or "Dam-fill" structure (hermetic adhesive + filling material) is used for this purpose.

3.3.5.3 Frit Encapsulation

Figure 3.39 shows the frit sealing structure [23]. Glass frit, with 8–20 µ height and 0.5–5.0 mm width, is dispensed on the cover glass as shown in the left side of Figure 3.40. Then, cover glass is attached on the OLED substrate and the frit portion is cured by moving laser irradiation. The frit glass reaches softening point by laser and is cooled down to make hermetic sealing of the device.

With frit sealing method, frit hermetic seal basically has very high barrier property, so very long reliability can be obtained for glass-substrate-based OLED devices.

3.3.5.4 WVTR Measurement

As discussed, a barrier layer for an OLED needs to provide a high protection capability against moisture and oxygen. Therefore, it is important to estimate the permeability of the device to ensure the OLED display product's lifetime. The most frequently used method to estimate the barrier property is the calcium test (Figure 3.41). Calcium is reactive to water and oxygen. These reactions transform calcium, an opaque electrically conductive metal, to CaO or Ca(OH)$_2$, a transparent insulator, which can be measured optically and electrically so that the vapor transmission across the barrier layer can be estimated [24].

Though the calcium test can be applied very easily and gives good accuracy, it does not discriminate between oxygen and water permeation. To

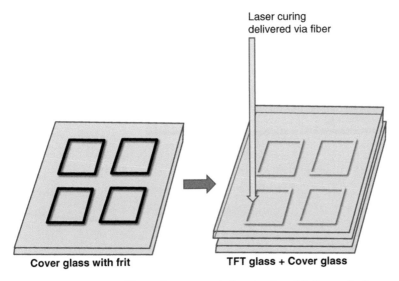

Cover glass with frit **TFT glass + Cover glass**

Figure 3.40 Examples of frit sealing encapsulation structure (bird's-eye view).

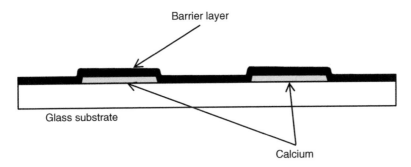

Figure 3.41 An example of sample structure for the calcium test.

get independent information about oxygen and water permeability, barrier property measurements are made using a vacuum system. Figure 3.42 shows the equipment used for the so-called Mocon method, for humidity transmittance measurement. Hundred percent humidity and low-humidity areas are separated by the barrier substrate, and the humidity transmitted through the barrier layer is detected by mass spectrometry. The detection limit of oxygen transmission is around 1×10^{-6} g/m^2·day with this method; however, it is hard to achieve a similar detection limit for water vapor transmission by the same method [24]. To make a better WVTR barrier property

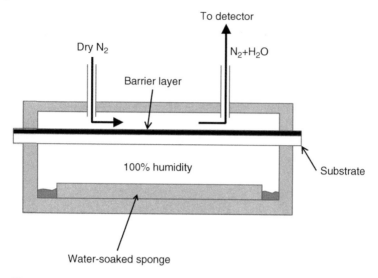

Figure 3.42 Equipment for the Mocon method of humidity permeation measurement [25].

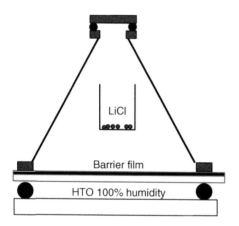

Figure 3.43 An example of HTO humidity measurement [26].

determination, the hydrogen–tritium–oxygen (HTO) method can be used. The HTO measurement uses similar equipment to the Mocon method but uses different moisture source and detector. The method uses HTO as the source of tritium and measures the tritium transmission rate (TTR). Figure 3.43 shows an example of barrier property measurement. The tritium transmitted through the barrier film is absorbed by LiCl. The LiCl is then dissolved in a scintillation cocktail, such as Perkin Elmer Ultima Gold LLT, and is measured by a scintillation counter [27]. The detection limit of the HTO method is below 1×10^{-6} g/m^2·day and is suitable for use in barrier property measurement for OLED devices.

3.4 PROBLEM ANALYSIS

When the properties of an OLED device seem to be defective, it is necessary to analyze the components of the device itself or the material remaining in the crucible to determine the cause of the problem. Methods used to troubleshoot and analyze the problem are discussed in the following sections.

3.4.1 Ionization Potential Measurement

As discussed in Section 2.3.6, it is very important to match the electrode work-function to the energy level of organic materials (HOMO and LUMO) to obtain high efficiency and low-voltage emission of the OLED device. An ionization potential analyzer is used for this purpose.

An ionization potential analyzer measures the relationship between the wavenumber of incident UV light and the number of photoelectrons that are emitted from the sample surface under this irradiation. The x-intercept of this relationship gives the workfunction (i.e., the minimum energy required to move an electron to a point at infinity; see Figure 3.44) and the semiconductor's ionization potential (the energy required to remove an electron to form an ion, also called *ionization energy*).

In the case of a metal, when the UV light energy is lower than the workfunction energy, no photoelectrons are emitted, and when it is higher, photoelectrons are emitted. A semiconductor behaves similarly; only when the energy of UV light is higher than the ionization potential are photoelectrons emitted. Thus, an energy potential analyzer measures the workfunction and ionization potential by these procedures.

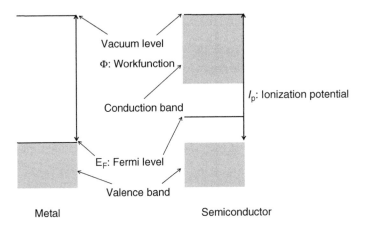

Figure 3.44 Schematic representation of workfunction and ionization potential.

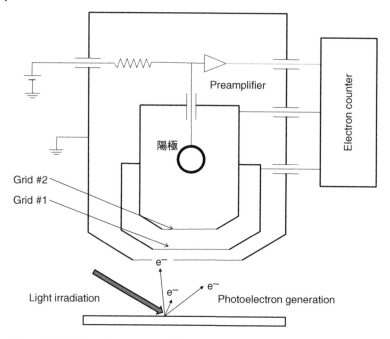

Figure 3.45 Schematic representation of operating principle of an ionization potential analyzer.

In the ionization potential analyzer shown in Figure 3.45, electrons emitted by the photoelectron effect due to UV light irradiation are captured by an oxygen molecule during their travel to the grid. Those electrons arrive at the anode after passing grids 1 and 2. When an ion approaches the anode, an electron avalanche occurs due to the presence of a strong electric field; one electron is amplified to approximately 10^5–10^7, and the discharge signal is provided to the preamplifier.

The number of electrons emitted depends on the energy of the irradiated light (Figure 3.46). The intercept of the graph in Figure 3.46 indicates the workfunction, and for the semiconductor, it gives the ionization potential, namely, HOMO energy level. This method is called photoemission spectroscopy (PES).

When an OLED device exhibits characteristics of fluctuation, or when a fabricated device does not manifest its intended characteristics, energy level measurement of the electrode and organic materials is very useful and sometimes leads to solution of the problem.

3.4.2 Electron Affinity Measurement

As discussed in Section 3.4.1, HOMO level can be measured using PES. However, it was not very easy to measure LUMO level until recently. To obtain LUMO level, three methods are reported [28].

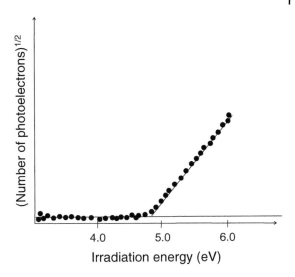

Figure 3.46 Graphical example of ionization potential analyzer measurement.

1. Reduction potential measurement by cyclic voltammetry (CV) method
2. Adding optical bandgap value to the HOMO level, which is measured by PES method
3. Inverse PES (IPES).

There were pros and cons for each measurement. CV method is carried out in solution, so it may not show accurate values. The second method can be carried out in solid phase; however, the optical bandgap measured by light absorption is often smaller than actual energy gap for an exciton.

IPES is an inverse process of PES method. PES method (Section 3.4.1) irradiates light onto the material and measure the outgoing electron. IPES process injects electron onto the sample and measures outgoing light. It can be carried out in solid phase. However, there are two major issues: measurement resolution and sample damage because the cross section of IPES is lower than that of PES by several orders of magnitude. Yoshida's method in *Chemical Physics Letters* [29] used low-energy electron; as a result, sample damage was avoided and also the resolution was dramatically improved because the outgoing light of Yoshida's method is near ultraviolet, different from the conventional IPES case, which used vacuum ultraviolet.

Due to the improvement of IPES method, direct LUMO level measurement has become possible without sample damage.

3.4.3 HPLC Analysis

High-performance liquid chromatography (HPLC; also called *high-speed liquid chromatography*) is frequently used to analyze impurities in OLED evaporation material and any residue material remaining in the crucible.

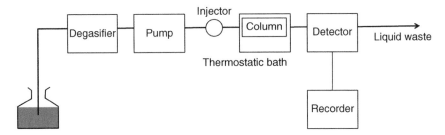

Figure 3.47 Flowchart of HPLC analysis equipment.

Figure 3.47 shows the equipment used for HPLC analysis. The typical components are a degasser, a pump, an injector, a thermostatic bath, a column, a detector, and a recorder.

First, the carrier solvent is transferred to the equipment from the carrier solvent bottle by a pump. (If pump pulsing occurs, a damper is sometimes used.) A degasser is necessary, as the presence of gases such as oxygen in the carrier solvent causes noise during detection when it is delivered to the equipment by a pump.

Next, liquefied samples are prepared for analysis by means of syringe injection and are delivered to the column. In the column, the samples are separated by partition between the flowing solvent and fine particles (e.g., 3–15 μm) of a stationary phase (often referred to as a gel; e.g., silica gel, polymer [polystyrene, polymethylmethacrylate, polyhydroxyethylmethacrylate, polyvinylalcohol, etc.] gel) supported in a stainless steel tube. As the separation process of the sample material is sensitive to temperature variation, the column is maintained in a thermostatic bath. The electrical signals provided by the detector after the separation process are detected and recorded (by the detector and recorder components, as shown in Figure 3.47).

There are many types of columns, such as reversed-phase columns, normal-phase columns, ion exchange columns, ligand exchange columns, ion exclusion columns, gas frontal chromatography (GFC) columns, and gel permeation chromatography (GPC) columns [30].

3.4.4 Cyclic Voltammetry

The cyclic voltammetry (CV) method, which measures potential and electric current simultaneously, is useful for analyzing electrochemical reactions by determining parameters such as electron transfer speed, reversibility, and chemical dissociation during reaction. Figure 3.48a shows a typical voltage sweep example for the measurement, and Figure 3.48b shows the corresponding electric current–voltage curve (cyclic voltammogram).

The electrode potential is controlled to ensure that the electrode is swept from a voltage higher than that of the oxidation–reduction potential to a lower

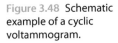

Figure 3.48 Schematic example of a cyclic voltammogram.

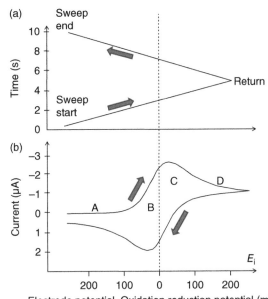

voltage and then swept back to the higher voltage again (see Figure 3.48a). In region *A*, there is almost no material to be oxidized, so the electric current is almost zero. In region *B*, the system undergoes reduction, so the electrons move toward the oxide, and the electric current thus rapidly increases.

In region *C*, when the voltage is slightly under the oxidation–reduction potential, the material near the electrode is almost completely reduced, so reduction no longer takes place, and the electric current starts to decrease. Gradually, the electric current created by reduction of material close to the electrode decreases, and the electric current is limited by reduction of material diffusing to the electrode from the bulk of the solution.

If the reduced material is stable, then when the voltage is swept from a lower value to a voltage higher than that of the oxidation–reduction potential, a reverse reaction takes place and the current–voltage curve shows complete charge symmetry; however, when the system is irreversible, a nonsymmetric *I–V* curve is created.

In some cases, this type of sweep is measured repeatedly. Also, dependence of the *I–V* curve on sweep speed is measured to enable analysis of the electron transfer mechanism in the submicrosecond range. This method provides information on the change in molecular structure with respect to electron transfer and electron mobility.

During the OLED operation, there occur surplus electrons or holes in the device system, which causes reduction or oxidation stress to organic molecules. To avoid such stress to cause reliability issue, tolerant molecule against such stress must be chosen. CV provides important insights to check such tolerance of molecules.

References

1 S. Son, S.-H. Yoon, J.-G. Jang, M.-S. Kang, and S.-Y. Jeon, Pure aluminum metal as an anode for high performance OLEDs, *SID 2004 Digest*, p. 52 (2004).

2 T. Yamada, T. Yoshimotod, and G. Ono, Organic EL devices using a novel hole-injection material, *Eurodisplay* 2002, p. 847 (**2002**).

3 J. Blochwitz, M. Pfeiffer, T. Fritz, and K. Leo, Low voltage organic light emitting diodes featuring doped phthalocyanine as hole transport material, *Appl. Phys. Lett.* **73**(6):729 (1998).

4 C. Adachi, M. A. Baldo, S. R. Forrest, and M. E. Thompson, Blue electroluminescence from phosphorescent benzoxazole doped OLEDs, *Proc. Materials Research Society* (1999 fall meeting), p. 12 (1999).

5 C. T. Brown, B. Owczarczyk, T. K. Hatwar, and D. Kondakov, High efficiency red emitting electroluminescent devices with exceptional stability, *IDRC 2003 Digest*, p. 42 (2003).

6 S. R. Forrest, M. A. Baldo, J. J. Brown, B. D'Andrade, Y. Kawamura, R. Kwong, M. E. Thompson, and S. Yanagida, Electrophosphorescent organic light emitting devices, *SID 2002 Digest*, p. 1357 (2002).

7 T. Nakayama, K. Hiyama, K. Furukawa, and H. Ohtani, Development of phosphorescent white OLED with extremely high power efficiency and long lifetime, *SID 2007 Digest*, p. 1018 (2007).

8 T. Tsuji, S. Kawakami, S. Miyaguchi, T. Naijo, T. Yuki, S. Matsuo, and H. Miyazaki, Red-phosphorescent OLEDs employing bis(8-quinolinolato)-phenolato-aluminum(III) complexes as emission-layer hosts, *SID 2004 Digest*, p. 900 (2004).

9 T. Tominaga, paper presented at 87th spring meeting, The Chemical Society of Japan, 2007, 1B1-38.

10 H. Kanno, Y. Hamada, K. Nishimura, K. Okumono, N. Saito, K. Mameno, and K. Shibata, Reduction in power consumption for full-color active matrix organic light-emitting devices, *Jpn. J. Appl. Phys.* **45**:L947–L950 (2006).

11 M. Ichikawa, T. Kawaguchi, K. Kobayashi, T. Miki, T. Obara, K. Furukawa, T. Koyama, and Y. Taniguchi, Bipyridyl oxadiazoles as a new class of durable and efficient electron-transporting materials, *SID 2005 Digest*, p. 1652 (2002).

12 T. Tohma, S. Yamazaki, and D. Wzorek, The future of active-matrix organic LEDs, *Information Display*, p. 20 (2001).

13 S. Van Slyke, A. Pignata, D. Freeman, N. Redden, D. Waters, H. Kikuchi, T. Negishi, H. Kanno, Y. Nishio, and M. Nakai, Linear source deposition of organic layers for full-color OLED, *SID 2002 Digest*, p. 886 (2002).

14 J. W. Hamer, A. Yamamoto, G. Rajeswaran, and S. A. Van Slyke, Mass production of full-color AMOLED displays, *SID 2005 Digest*, p. 1902 (2005).

15 S. F. Lim, L. Ke, W. Wang, and S. J. Chua, Correlation between dark spot growth and pinhole size in organic light-emitting diodes, *Appl. Phys. Lett.* **78**(15):2116 (2001).

16 Y.-F. Liew, H. Aziz, N.-X. Hu, H. S.-O. Chan, G. Xu, and Z. Popovic, Investigation of the sites of dark spots in organic light-emitting devices, *Appl. Phys. Lett.* **77**(17):2650 (2000).

17 M. Schaer, F. Nüesch, D. Berner, W. Leo, and L. Zuppiroli, Water vapor and oxygen degradation mechanisms in organic light emitting diodes, *Adv. Funct. Mater.* **11**(2) (2001).

18 G. Nisato, M. Bouten, L. Moro, O. Philips, and N. Rutherford, Thin film encapsulation for OLEDs: evaluation of multi-layer barriers using the Ca test, *SID 2003 Digest*, p. 88 (2003).

19 W. Prins, J. J. Hermans, *J. Phys. Chem.* **63**:716 (1959).

20 T. Tsujimura and T. Miyamoto, Wire breakdown method for evaluating metal taper shape, International Display Workshop 1997 Digest, p. 219 (1997).

21 D. H. Levy, S. F. Nelson, and D. Freeman, Oxide electronics by spatial atomic layer deposition, *IEEE/OSA J. Display Technol.* **5**(12):484–494 (2009).

22 S. Hong, C. Jeon, S. Song, J. Kim, J. Lee, D. Kim, S. Jeong, H. Nam, J. Lee, W. Yang, S. Park, Y. Tak, J. Ryu, C. Kim, B. Ahn and S. Yeo, Development of commercial flexible AMOLEDs, *SID 2014 Digest*, Vol. 38, Issue 1, pp.1701–1704 (2014).

23 L. Zhang, S. Logunov, K. Becken, M. Donovan, and B. Vaddi, Impacts of glass substrate and frit properties on sealing for OLED lighting, *SID 2010 Digest*, Vol. 41, Issue 1, pp. 1890–1893 (2010).

24 J.-S. Park, H. Chae, H. K. Chung, and S. I. Lee, Thin film encapsulation for flexible AM-OLED: a review, *Semiconductor Science and Technology* **26**(034001):1 (2011).

25 F. McCormick, Barrier Films & Device Encapsulation, Council for Chemical Research OLED Materials for Lighting and Displays Workshop Session 4 (2011).

26 M. D. Groner, S. M. George, R. S. McLean, and P. F. Carcia, Gas diffusion barriers on polymers using Al_2O_3 atomic layer deposition, *Appl. Phys. Lett.* **88**(051907):88 (2006).

27 A. A. Dameron, S. D. Davidson, B. B. Burton, P. F. Carcia, R. S. McLean, and S. M. George, Gas diffusion barriers on polymers using multiple layers fabricated by Al_2O_3 and rapid SiO_2 atomic layer deposition, *J. Phys. Chem. C* **112**:4573–4580 (2008).

28 H. Yoshida and K. Yoshizaki, Electron affinities of organic materials used for organic light emitting diodes: a low-energy inverse photoemission study, *Org. Electron.* **20**, 24 (2015).

29 H. Yoshida, Near-ultraviolet inverse photoemission spectroscopy using ultra-low energy electrons, *Chem. Phys. Lett.* **539–540**, 180–185 (2012).

30 S. Denko, Distribution measurement and analysis of additives by GPC solvent separation column. *Shodex Chromato News* **70** (1998).

4

OLED Display Module

4.1 COMPARISON BETWEEN OLED AND LCD MODULES

Figure 4.1 compares the components that are necessary for production of liquid crystal display (LCD) and OLED display modules.

An LCD consists of many components because it must convert backlight emission to uniform area emission and switch on and off the light with a liquid crystal shutter, which is positioned between two polarizers.

A typical LCD uses multiple LEDs or a cold-cathode fluorescent tube (CCFL). Two types of LED are used: (1) that emits short wavelength emission, which is converted to longer wavelengths by means of a fluorescent material (downconversion) and (2) that emits the three color primaries (red–green–blue [RGB]). Thus, the light source for an LCD is either linear (fluorescent tube) or point, so the light must be converted to the area form to be used as a backlight unit. The light reflected by the reflector passes through the light guide and is diffused. A light guide is made of high refractive index material, such as an acrylic polymer, which delivers the light by total internal reflection due to the refractive index difference between it and the surrounding air. The light guide structure is designed such that uniform luminance distribution across the area of the display can be achieved. Light exiting from the light guide is transmitted through the prism sheet and diffuser and is then polarized by the rear polarizer. The polarization state of the light is altered by the birefringent LC layer and controlled by the voltage applied to it. The RGB components of the light are selected by a color filter aligned with the primary color subpixels of the display, and the exit polarizer transmits the required intensity of light—according to its polarization—toward the viewer.

In OLED displays, light emission takes place in an OLED device fabricated on the surface of a glass substrate. When the display is observed in a bright environment, the reflection from the display surface causes deterioration of

OLED Display Fundamentals and Applications, Second Edition. Takatoshi Tsujimura.
© 2017 John Wiley & Sons, Inc. Published 2017 by John Wiley & Sons, Inc.

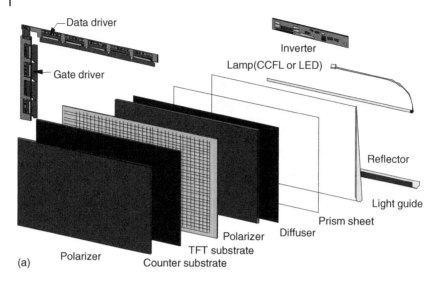

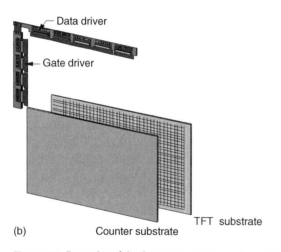

Figure 4.1 Examples of display components used in LCD (a) and OLED (b) television sets.

the contrast ratio, so a circular polarizer is often used (see Section 4.4.3 for further details), which reduces the luminance of the display by about 50%. For example, a 45% transmittance polarizer is sometimes used. LCDs also have an ambient light reflection problem. In an LCD, a black matrix is used to enhance the ambient contrast ratio. Generally, the reflection is more serious in the case of OLED because reflective electrode is used for OLED when transmissive LCD uses transparent electrode for both sides.

In all cases, OLED displays require fewer components than LCDs. O'Regan et al. [1] report the cost structure comparison between LCD and OLED. The paper ascribes the major reason of OLED's high cost to the low material utilization, which is typically 5% by evaporation methods. Even assuming a significantly higher cost of TFT backplane and driver IC than LCDs and also assuming a lower yield of OLED than LCD by 10–20%, the total cost of OLED would become 30% cheaper if a high material utilization processing technique is used, such as a solution-based technique. Recently, it was reported by a display research company that 5-in. full-HD AMOLED display cost became lower compared to that of a similar LCD display [2]. OLED TV is also showing significant cost reduction these days due to yield improvement by white + color filter architecture and also due to the adoption of large mother glass size. It is reported that the display research company is expecting OLED TV to reach a premium of 10% over LCDs [3].

4.2 BASIC DISPLAY DESIGN AND RELATED CHARACTERISTICS

To design a high-quality display, many factors must be considered. Designers need to understand how intended specifications, such as luminance, color reproduction, and luminance uniformity, can be achieved through display design.

4.2.1 Luminous Intensity, Luminance, and Illuminance

The radiometric (energy) values taking spatial and temporal considerations into account are called radiant quantities, and radiometric values weighted by physiological functions related to the wavelength sensitivity of the eye (spectral luminous efficacy) are called photometric quantities. Luminous intensity, luminance, and illuminance are the most well-known photometric quantity units.

4.2.1.1 Luminous Intensity

Here we assume that the luminous flux from a point light source traverses the areas between dm_1 and dm_2 as shown in Figure 4.2. Let us assume that the luminous intensity values are P_1 and P_2. The luminous flux values Φ (in lumens [lm]) of areas dm_1 and dm_2 are the same:

$$d\Phi = P_1 dm_1 = P_2 dm_2 \tag{4.1}$$

If the solid angle is $d\omega$, then the luminous flux per unit solid angle P is

$$P = \frac{d\Phi}{d\omega} \tag{4.2}$$

This parameter is called *luminous intensity*, measured in cd (candelas) = lm/sr (where sr is steradian, the unit for solid-angle measurement).

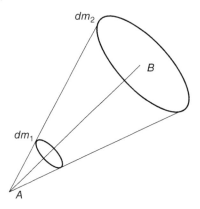

Figure 4.2 Relationship between luminous flux and luminous intensity.

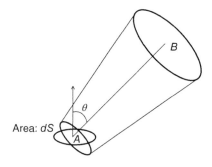

Figure 4.3 Relationship between luminous flux and luminance of an area light source.

4.2.1.2 Luminance

Luminous intensity can be measured in terms of a point light source, but measurement of an area light source requires another metric. As an area light source can be quantified as a multipoint light source, the luminous flux of an area light source is much larger than that of a point light source. The luminous intensity of the area light source is called *luminance*, and the measurement unit is $cd/m^2 = lm/(sr \cdot m^2)$. ($cd/m^2$ is sometimes called "nit," but it is not an officially acknowledged unit name.) As shown in Figure 4.3, when the angle between the normal line of face A and the $A–B$ line is θ, the area of face A is regarded as $dS \cos \theta$ from the location on line $A–B$, so the luminance L can be expressed as follows:

$$L = \frac{dP}{dS \cos \theta} \tag{4.3}$$

Here, using 4.2, we obtain

$$L = \frac{d\Phi}{dS \cos \theta \, d\omega} \tag{4.4}$$

It should be noted that the luminance and the brightness are not the same (discussed in Section 4.2.1.5).

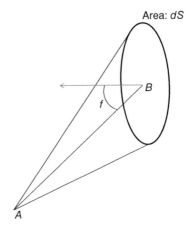

Figure 4.4 Relationship between luminous flux and illuminance of point light.

4.2.1.3 Illuminance

In Section 4.2.1.2, we noted that the luminous flux emitted per unit solid angle is called *luminous intensity*. Similarly, the luminous flux incident per unit area is called *illuminance*, and the measurement unit is lux (lx) ($1\,lx = 1\,lm/m^2$). We can use the same equation (Eq. 4.4) to calculate the *luminous exitance*, which is the luminous flux that exits from the source per unit area.

Illuminance E (luminous exitance) can be expressed as

$$E = \frac{d\Phi}{dS} \tag{4.5}$$

where, as shown in Figure 4.4, dS is the area on the incident face of luminous flux $d\Phi$.

Now let us consider how illuminance E of face B is defined. If the angle between face B and line A–B is ϕ, then the area of face B can be regarded as $dS \cos \phi$ from line A–B. If the distance between points A and B is d_{AB}, the solid angle $d\omega$ that is occupied by this space can be expressed as follows:

$$d\omega = \frac{dS \cos \phi}{d^2_{AB}} \tag{4.6}$$

For the point light source (Figure 4.4), using Eq. (4.2), we obtain

$$d\Phi = \frac{P dS \cos \phi}{d^2_{AB}} \tag{4.7}$$

Therefore, the illuminance E for a point light source is

$$E = \frac{d\Phi}{dS} = \frac{P \cos \phi}{d^2_{AB}} \tag{4.8}$$

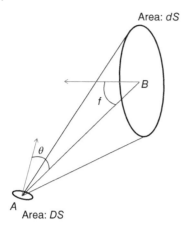

Area: *dS*

Figure 4.5 Relationship between luminous flux and illuminance of area light.

For the area light source (Figure 4.5), the area denoted by ΔS, the following formula can be derived using ΔS instead of dS in (Eq. 4.4):

$$d\Phi = L \Delta S \cos\theta \, d\omega$$
$$= \frac{L \Delta S \cos\theta \, dS \cos\phi}{d_{AB}^2} \tag{4.9}$$

Therefore, the illuminance of an area light source is

$$E = \frac{d\Phi}{dS}$$
$$= \frac{L \Delta S \cos\theta \cos\phi}{d_{AB}^2}. \tag{4.10}$$

Illuminance is the parameter used to quantify the brightness of an environment.

4.2.1.4 Metrics Summary

The four quantities discussed in the preceding section can be summarized as follows:

- *Luminous flux* is the psychophysical (in terms of spectral sensitivity of observers' eyes) visible light intensity per unit time, emitted by a light source.
- *Luminous intensity* is the luminous flux emitted per unit solid angle.
- *Luminance* is the luminous flux per unit solid angle emitted per unit area as projected on a plane normal to the line of sight.
- *Illuminance* is the luminous flux incident per unit area on a surface.

The equations and units used to calculate these quantities are summarized in Table 4.1.

Table 4.1 Summary of Photometric Parameters

Parameter	Quantification Equation	Unit
Luminous flux	$\Phi = \dfrac{dQ}{dt}$	Lumen (lm)
Luminous intensity	$P = \dfrac{d\phi}{d\omega}$	Candela $=$ lumen/steradian $(cd = lm/sr)$
Luminance	$L = \dfrac{d\Phi}{dS \cos\theta \, d\omega}$	Candela per square meter $=$ lumen per steradian per square meter $(cd/m^2 = lm/[sr \cdot m^2])$
Illuminance	$E = \dfrac{d\Phi}{dS}$	Lux $(lx = lm/m^2)$

The steradian (sr) is the unit for determining solid angle. (A steradian means the solid angle that occupies the area equal to the square of the spheric radius, on the sphere surface. The solid angle for a full sphere is 4π.) When the surface reflection of a specific display is discussed, illuminance and luminance are often compared to calculate the contrast ratio in a lit environment as follows.

If perfect diffusion is assumed for surface A, illuminance at point B can be expressed as

$$E_B = \frac{dI(\theta) \cdot \cos^2\theta}{(2r \cdot \cos\theta)^2} = \frac{dI(\theta)}{4r^2}$$

As the illuminance is uniform at all points on the surface of the circle with radius r, total luminous flux can be calculated as

$$\Phi = \frac{I_0 \cdot 4\pi r^2}{4r^2} = \pi \cdot I_0$$

As surface A is perfectly diffusive, luminance of reflective light on surface A can be described as

$$L_{perfect} = \frac{E}{\pi}$$

E is the illuminance on surface A.

If reflectivity of surface A is K, this equation can be written as

$$L_{reflection} = \frac{K \cdot E}{\pi}$$

The living room contrast ratio, discussed in Section 4.4.3.1, can be calculated by

$$CR_{livingroom} = \frac{L_{display}}{L_{reflection}} = \frac{\pi \cdot L_{display}}{K \cdot E}$$

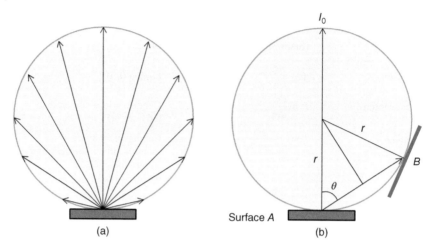

Figure 4.6 (a) Perfect diffusion surface; (b) incident light from a perfect diffusion surface. (Lambertian distribution.)

Normally, $E = 200$ (lx) is used for living room contrast ratio calculation. See Figure 4.6.

4.2.1.5 Helmholtz–Kohlrausch Effect

It is known that visual perception is subjected to the viewing condition. For example, when the colors are kept the same and the ambient light is changed, the perceived brightness and color change (Purkinje effect) [4]. Similarly, it has been reported that the characteristics of OLED display, such as high contrast and large color gamut, is affecting the perceived performance of the display.

It is known that perceived brightness is increased as the color is saturated. This is called Helmholtz–Kohlrausch effect (also called "H–K effect") [5, 6]. As many modern displays have a large color gamut, it can be perceived as brighter [7]. It is known that darker backgrounds makes human feel brighter (Bartle-son–Breneman effect [8]). Here, it should be noted that brightness and luminance have different meaning. Brightness is a subjective attribute of light, which human perceives. On the other hand, luminance is luminous intensity per unit area projected in a given direction. It is reported that, due to psychological effect, OLED display's brightness can be perceived 25% higher than LCDs with the same measured luminance [9].

4.2.2 OLED Current Efficiencies and Power Efficacies

When multiple OLED devices or OLED processes are compared or controlled, it is much easier to compare the efficiency metric than to compare actual measurement curves with many data points. The concepts of current efficiency and power efficacy (luminous efficacy) are discussed as follows.

(Readers should note that there is an ongoing discussion whether "efficiency" or "efficacy" should be used for this topic. Standardization committee, especially for lighting field, such as in IEC: International Electrotechnical Commission and in DOE: Department Of Energy, are promoting the use of "luminous efficacy." The main reason is that the word "efficiency" should be dimensionless and the word "efficacy" can carry the meaning of physiological quantity such as the eye response curve. "Current efficiency/efficacy" is also under discussion, but still no clear answer is yet. So, in this book, the author uses "luminous efficacy," "power efficacy," and "current efficiency" throughout.)

Let us assume monochromatic light first. The luminous flux $\Phi(\lambda)$ (measured in lm/m^2) at wavelength λ can be expressed as

$$\Phi(\lambda) = PK_m y(\lambda) \tag{4.11}$$

where P is emission intensity (W/m^2), $y(\lambda)$ is the relative luminosity at wavelength λ, and K_m is the maximum luminosity factor 683 (lm/W). An actual emission has spectral width, so an integral that can be applied for all wavelengths is

$$\Phi(\lambda') = PK_m y(\lambda') \frac{F(\lambda')}{\int F(\lambda)d\lambda} \tag{4.12}$$

Luminous flux Φ for all wavelengths can be calculated by integrating for all λ' as follows:

$$\Phi = PK_m \frac{\int F(\lambda)y(\lambda)d\lambda}{\int F(\lambda)d\lambda} \tag{4.13}$$

Then the luminance L (cd/m^2) can be expressed as

$$L = \frac{\Phi}{\pi} \tag{4.14}$$

The power efficacy (lm/W; also called luminous efficacy) can be expressed as

$$\eta_e = \frac{\Phi}{W} = \frac{PK_m y(\lambda)}{S} \cdot \frac{S}{IV} = \frac{PK_m y(\lambda)}{IV} \tag{4.15}$$

On the other hand, current efficiency η_c (cd/A) is defined as the luminance divided by the current density:

$$\eta_c = \frac{L}{J} = \frac{PK_m y(\lambda)}{\pi \cdot I} \tag{4.16}$$

An actual current–voltage (*I–V*) curve of an OLED device is illustrated in Figure 4.7, which shows that the electric current flows when the voltage is positive but does not when it is negative, which means that the OLED device

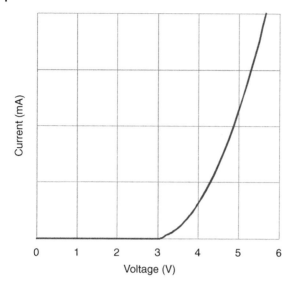

Figure 4.7 Example of a current–voltage curve of an OLED device.

exhibits diode characteristics. For nonuniformity compensation of a display (described in Section 6.4.4), many pixel circuitry requires to stop the OLED current flow while the threshold voltage of driver TFT is measured, so an additional TFT serially connected to OLED was used in the past. The diode feature of OLED can be utilized to remove the additional TFT, as it impedes the current flow of the driver TFT by applying negative voltage, which obstructs the detection of the driver TFT characteristic fluctuation to be compensated by a pixel circuitry.

A luminance–voltage curve is shown in Figure 4.8. This curve is also called the $L–V$ curve. Similarly, luminance–current–voltage measurement is sometimes called $L–I–V$ measurement. (It should be noted that L means luminance here, not inductance.)

As shown in Figure 4.9, the luminance is almost proportional to the electric current. The correlation between current efficiency and voltage is plotted in Figure 4.10. Thus, the efficiency is subjected to the voltage (due to concentration quenching, described in Section 2.4.2.4; densely generated excitons by larger current cause more chance of losing energy by collision to the quencher or other excitons), which needs to be taken into account in the display design. In particular, in the case of a phosphorescent device, collision between two triplet excitons causes quenching (as the high current generates many excitons by which many triplet excitons collide with each other to form a singlet state). This is called *triplet–triplet (T–T) annihilation* (discussed in Section 2.4.2.4), which can significantly reduce efficiency in the high-current region. This efficiency reduction is referred to as the *rolloff phenomenon*.

Figure 4.8 Example of a
luminance–voltage curve
of an OLED device.

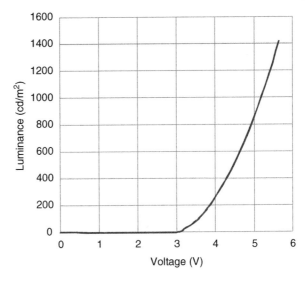

Figure 4.9 Example plot of
luminance versus current.

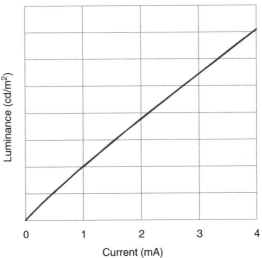

4.2.3   Color Reproduction

Color reproduction is an important factor in ensuring color fidelity, especially
for photo images. Large-scale color reproduction can be achieved by applying
display schemes containing the primary colors red, green, and blue, as discussed below. Emission spectra of red, green, and blue subpixels are shown in
Figure 4.11.

Here we assume that the emission spectrum is $S(\lambda)$ and the color matching
functions (sensitivity of the human eye to each wavelength; see the appendix for

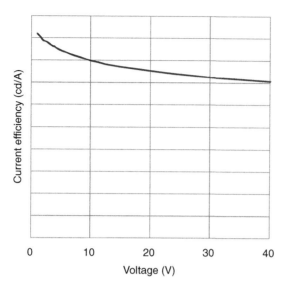

Figure 4.10 Graph showing dependence of voltage on current efficiency.

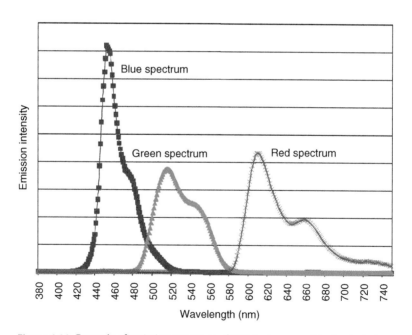

Figure 4.11 Example of emission spectrum of RGB (red+green+blue) subpixels.

actual values) are $\bar{x}(\lambda)$, $\bar{y}(\lambda)$, $\bar{z}(\lambda)$; we can then express the tristimulus values (a parameter set, derived from human eye responses to three color primaries, designed so that all weight factors are positive; standardized as the CIE1931 color space) as follows, using an integral ranging from 380 to 780 nm.

$$X = K \int_{380}^{780} S(\lambda)\bar{x}(\lambda)d\lambda \tag{4.17}$$

$$Y = K \int_{380}^{780} S(\lambda)\bar{y}(\lambda)d\lambda \tag{4.18}$$

$$Z = K \int_{380}^{780} S(\lambda)\bar{z}(\lambda)d\lambda \tag{4.19}$$

Here, we define the K so that calculated Y is equal to the display luminance.

Using the CIE1931-XYZ value, the (x, y) locus (generally used to indicate the display color reproduction capability) can be expressed as follows:

$$x = \frac{X}{X + Y + Z} \tag{4.20}$$

$$y = \frac{Y}{X + Y + Z} \tag{4.21}$$

This (x, y) value set can be calculated for red, green, and blue, respectively, as (x_R, y_R), (x_G, y_G), and (x_B, y_B).

The actual measurement results of an OLED emitter are shown in Figure 4.12. The color range that can be shown by the combination of these three color primaries is found within the triangle that is made by the three color primaries in the (x, y) coordinate. This area is called the *color gamut*. To determine the color reproduction capability of a display, the metric NTSC% is often used (NTSC = National [US] Television System Committee) (Figure 4.13). It is defined as the ratio of the area of the triangle achievable by the display divided by the area of the standard NTSC triangle, expressed as a percentage. For example, XEL-1, the 11-in. OLED television commercialized in 2007, claimed a color gamut of 110%.

Although NTSC was introduced as a cathode ray tube (CRT) television standard, eventually most CRT fluorescence materials were designed with emission color coordinates close to the European Broadcasting Union (EBU) standard or the Rec709 standard (the international standard for HDTV studios), which are much smaller in area than that of the NTSC triangle (Figure 4.14).

On the other hand, digital cameras and computers often use the s-RGB standard, which is shown in Figure 4.14 (it has the same color coordinates as EBU and Rec709). The Joint Photographic Experts Group (JPEG) standard also normally uses the s-RGB color coordinate standard; however, as the s-RGB triangle is not large enough to reproduce high fidelity colors, a new header standard

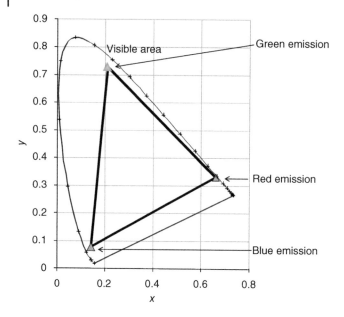

Figure 4.12 Plot of the area reproducible by RGB color mixture.

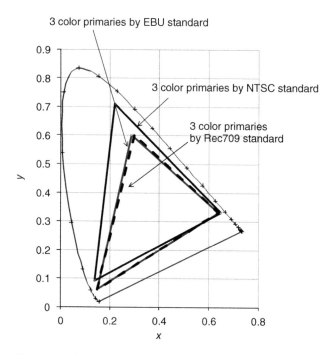

Figure 4.13 Graphical representation of the three color standards applied in television manufacture.

Figure 4.14 Graphical representation of the three color standards applied in digital camera manufacture.

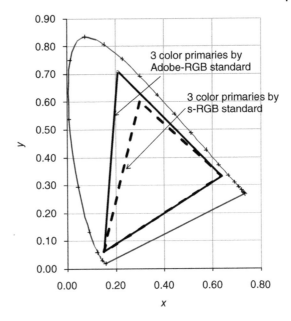

EXIF2.2 has introduced the s-Ycc standard, which uses an extended dynamic range with grayscale values that extend from negative values to values greater than 255, while normal s-RGB uses 0–255 to express grayscales. The newer header standard EXIF2.21 can also handle the Adobe-RGB standard, which is popular in the publishing industry (Figure 4.14).

For better user experience, color reproduction enhancement of displays is attracting attention. Especially for the latest "Super Hi-Vision" standard, very wide color description BT-2020 is applied, which covers 99% of the so-called, "Pointer's color" region, which was introduced by Pointer that describes the area of surface object colors in real world [10]. DCI-P3 standard intended for digital movie projection is also becoming prominent, which covers most of the Pointer's color regions.

The (x, y) color coordinates for four of these standards are listed in Table 4.2.

To ensure color fidelity, s-RGB can reproduce most colors realistically because there are not many vivid colors outside the s-RGB triangle in the real world, as illustrated in Figures 7.2 and 7.3. However, although highly saturated colors rarely occur, they have an intense impact on a viewer, so a color gamut wider than that of s-RGB is useful for showing such colors.

In the television industry, the image boosting technique, which converts an original image into a more impressive image, such as one with more vivid colors, by means of graphic engine IC chips or a graphics processing unit (GPU), is often used to enhance the image impression. A wider color gamut is useful for such an application as well (discussed in Section 4.2.6).

Table 4.2 (x, y) Coordinate of Color Primaries for Each Standard

Standard	Red	Green	Blue
NTSC	$(0.670, 0.330)$	$(0.210, 0.710)$	$(0.140, 0.080)$
Rec-709 (HDTV)	$(0.640, 0.330)$	$(0.300, 0.600)$	$(0.150, 0.060)$
EBU	$(0.640, 0.330)$	$(0.290, 0.600)$	$(0.150, 0.060)$
s-RGB	$(0.640, 0.330)$	$(0.300, 0.600)$	$(0.150, 0.060)$
Adobe-RGB	$(0.640, 0.330)$	$(0.210, 0.710)$	$(0.150, 0.060)$
BT-2020	$(0.708, 0.292)$	$(0.170, 0.797)$	$(0.131, 0.046)$
DCI-P3	$(0.680, 0.320)$	$(0.265, 0.690)$	$(0.150, 0.060)$

As discussed earlier, the (x, y) coordinate is popular as a display color metric; however, there are problems with using the standard. Distance in the (x, y) coordinate is not proportional to differences in perception of the human visual system, so a larger area does not always mean better display capability. Figure 4.15 shows 10 times the perception limits of the human eye in different regions of the (x, y) color space reported by McAdam et al. [11]; this is known as the "McAdam ellipse." As it is an ellipse, not a circle, the human eye does not have the same sensitivity for the x axis and the y axis. Also as shown in Figure 4.15, the human eye is very sensitive to blue colors (left bottom of the graph) but is less sensitive to red (right bottom) and is very insensitive to green colors (top). This illustrates that human eye sensitivity depends on the location on the (x, y) coordinate. Therefore, it is not very meaningful to discuss the area to judge the display capability on the basis of an (x, y) coordinate system.

The issues discussed above explain why many display companies and research institutes have used (u', v') color coordinates to express display capability. The CIE1976 color space coordinates (u', v') can be expressed as

$$u' = \frac{4x}{(-2x + 12y + 3)} = \frac{4X}{(X + 15Y + 3Z)} \tag{4.22}$$

$$v' = \frac{9y}{(-2x + 12y + 3)} = \frac{9Y}{(X + 15Y + 3Z)}. \tag{4.23}$$

Performance comparisons using the (u', v') color coordinates indicate how a color display color can actually be perceived by the human eye. For this reason, there is an increasing number of studies in the literature, calculating color gamut by (u', v').

Also, uniform color space is sometimes factored into any discussion of display capability because the defective criteria (such as the luminance variation limit or the image burning limit according to lifetime test discussed in

Figure 4.15 Graphical representation of McAdam ellipse, showing the human eye's perception limit (at 10° magnification).

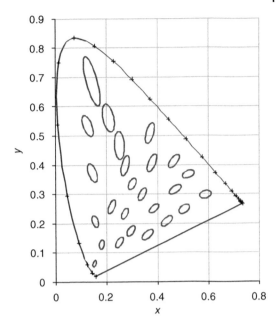

Section 2.5, uniformity criteria discussed in Section 6.4.4, or the metal bus design discussed in Section 6.4.3) should be determined according to the limit of human perception.

4.2.4 Uniform Color Space

CIE-LAB and CIE-UV are two well-known uniform color spaces. On these uniform color spaces, lightness and color can be treated equally so that the length $=1$ on the color space is almost equal to the human eye perception limit.

In CIE-LAB uniform color space (Figure 4.16), lightness L^* is defined as

$$L^* = 116f\left(\frac{Y}{Y_n}\right) - 16$$

$$a^* = 500\left\{f\left(\frac{X}{X_n}\right) - f\left(\frac{Y}{Y_n}\right)\right\}$$

$$b^* = 200\left\{f\left(\frac{Y}{Y_n}\right) - f\left(\frac{Z}{Z_n}\right)\right\}$$

In the case of $t > \left(\frac{6}{29}\right)^3$, $f(t) = t^{1/3}$. In other cases, $f(t)$ is $\frac{1}{3}\left(\frac{29}{6}\right)^2 t + \frac{4}{29}$.

Plotting a^*, b^*, and L^* on a three-dimensional graph, the angle of the line made by the plotted point and the original point in plane a^*–b^* shows the hue, and the distance between the plotted point and the original point in plane a^*–b^*

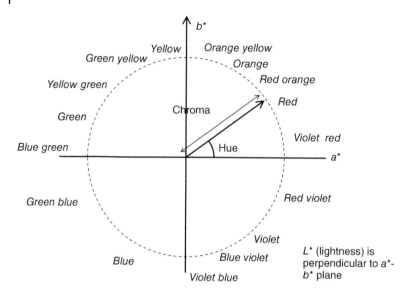

Figure 4.16 CIE-LAB uniform color space.

is the chroma, as in the Munsell color system widely used in color science. In this color space, the color difference (including lightness difference) is defined as ΔE_{ab} by the following equation:

$$\Delta E_{ab} = \sqrt{\Delta L^{*2} + \Delta a^{*2} + \Delta b^{*2}}$$

For CIE-LUV uniform color space, the (u^*, v^*) coordinate is defined by the following equations:

$$u^* = 13L^*(u - u'_n)$$
$$v^* = 13L^*(v - v'_n)$$

Color difference can be expressed for CIE-LUV uniform color space as follows:

$$\Delta E_{uv} = \sqrt{\Delta L^{*2} + \Delta u^{*2} + \Delta v^{*2}}$$

Color difference $\Delta E = 1$ is known to be close to the human perception limit and is a useful criterion for determining parameters for display design, such as

- Acceptable display white point variation
- Acceptable lifetime and viewing angle
- Acceptable image sticking level.

4.2.5 White Point Determination

To display a white image, all red, green, and blue subpixels need to emit light. The white color changes the impression of a display quite a bit. The color should be selected according to the purpose of the display.

When tristimulus values of red, green, and blue are (X_R, Y_R, Z_R), (X_G, Y_G, Z_G), (X_B, Y_B, Z_B) respectively, then

$$X_{white} = X_R + X_G + X_B \tag{4.24}$$

$$Y_{white} = Y_R + Y_G + Y_B \tag{4.25}$$

$$Z_{white} = Z_R + Z_G + Z_B \tag{4.26}$$

Then, the (x, y) coordinate of the white emission can be calculated as follows:

$$x_{white} = \frac{X_{white}}{X_{white} + Y_{white} + Z_{white}} \tag{4.27}$$

$$y_{white} = \frac{Y_{white}}{X_{white} + Y_{white} + Z_{white}}. \tag{4.28}$$

The white coordinate (white point) is expressed by the CIE coordinates as well as by the blackbody radiation temperature.

To express the color of white emission, a blackbody temperature is often used. Radiant intensity from a black object having temperature T at the wavelength between λ and $d\lambda$ can be expressed as (Planck radiation law)

$$P(\lambda) = \frac{8\pi hc^2}{\lambda^5} \cdot \frac{1}{e^{\frac{hc}{\lambda k_B T}} - 1}$$

(P, total radiant intensity from a black object toward all directions; c, speed of light; h, Planck constant; k_B, Boltzmann constant. It is necessary to pay attention to a potentially confusing definition. "Warm" color temperature, which normally means a color coordinate closer to red, has a low color temperature, while "cold" color temperature, normally a color coordinate closer to blue, has a high color temperature.)

Figure 4.17 shows the spectrum of the radiant intensity for various color temperatures.

Figure 4.18 shows the CIE coordinate of each blackbody temperature. (The line that the color coordinate follows when temperature is changed is called Planckian locus or blackbody locus.)

In television applications, a temperature of 6500 K (designated as D65) is used to display the white point, which has been introduced by the SMPTE-170M standard in the United States and many other countries. On the other hand, Asian countries such as Japan are using high-temperature color points. For example, Japan is using 9300 K (D93) as a target, which was introduced by NHK, a Japanese broadcasting organization. D65 is $(x, y) = (0.3127, 0.3290)$ and D93 is $(x, y) = (0.283, 0.297)$ on the CIE coordinate. For computer monitors, three color point temperatures are sometimes available: 5000, 6500, and 9300 K.

When designing OLED displays, it is necessary to adjust the red, green, and blue primary color intensities to ensure adequate white balance.

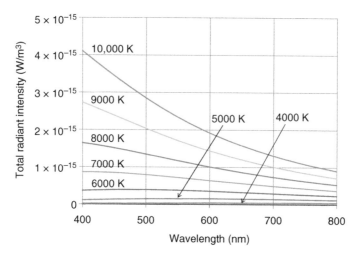

Figure 4.17 Blackbody emission spectrum for each color temperature.

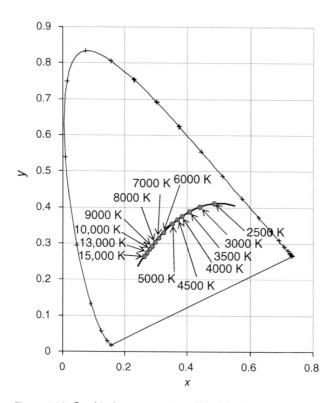

Figure 4.18 Graphical representation of blackbody versus color temperatures.

4.2.6 Color Boost

As discussed in Section 4.2.3, a trend exists to aim for wider color reproduction to enhance the user's viewing experience. There was a case that achieved 190% color gamut compared to Rec-709 standard by means of laser scanning television [12]. Then the question arises whether a television with a wider color gamut is always better.

For fidelity, if a television system is using EBU standard, wider color reproduction than EBU color point does not make sense. A still image on PC using s-RGB standard with zero to 255 values (which is representing grayscales by 8-bit, namely, $2^8 = 256$ colors) does not need a wider gamut for fidelity color reproduction. As discussed in Section 7.4.1.2, as the color is saturated (as the coordinate goes the outer direction), the probability to have the color point is significantly decreased, especially outside the EBU triangle.

The reasons why people want to use a wider color gamut is one of the following cases or both.

(1) A display that needs to reproduce rare saturated colors
(2) A display with color boosting for better viewer experience.

In option (1) without color boosting, the viewers seldom see the color outside the EBU triangle. It should be noted though such rare colors seldom appear in real world, it remains much stronger impression than frequent colors. The purpose of having a wider color gamut in such systems is to have impressive color reproduction for rare images.

Option (2) is a more realistic implementation of wider color reproduction. It is using human's preference of converted object colors to more vivid images. Hisatake et al. made an analysis on human's preference on the color conversion [13]. Figure 4.19 shows the allowable limit and optimal limit of color conversion. In most cases, color conversion to less vivid direction (inward conversion in the figure) is not acceptable, but conversion to more vivid direction is more acceptable (outward direction). It was also pointed out that wider color conversion within the allowable limit is preferred by the viewers. (It can be also seen that the color conversion near the skin color is not preferred, very much different from other colors.) From the result, a certain level of color conversion to more vivid colors within allowable limit improves the viewing experience. Wider color gamut display helps to make this happen.

Generally speaking, a display with wider color reproduction delivers better viewing experience through color boosting and the capability to show rare colors. Said that, it should be noted that the effectiveness is limited beyond 100%

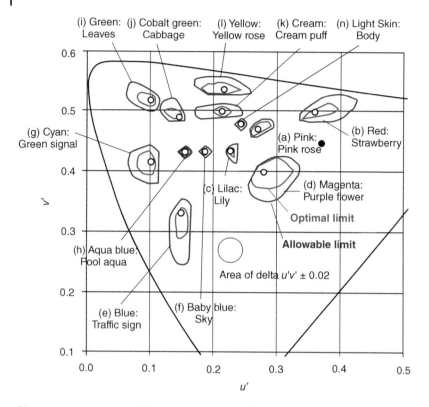

Figure 4.19 Optimum and limit of color boosting (Courtesy by Y. Hisatake).

NTSC. At the same time, as the vivid primary colors have less human eye's sensitivity, the power consumption is increased as the display color gamut is increased. It is reasonable to design an OLED display taking the trade-off into account.

4.2.7 Viewing Condition

Kubota et al. made a survey analysis on the viewing condition of CRT display in home use environment for 50 Japanese households [14]. The data shows the following:

1. Illuminance on the display was 27–327 lx and average was 108 lx.
2. Luminance of display was 72–530 cd/m², average was 260 cd/m².
3. Viewing angle was ±60° for horizontal direction and was ±30° for vertical direction. Moriguchi et al. reported that there is an acceptable viewing angle limit due to the geometrical distortion of a display [15].

Tanton et al. also made a survey in England by 102 people regarding the viewing distance in 2004. The mode value of 50 households was 8H, 8 times the display screen height [16]. Noland et al. reported that the display size is increasing without much change in the viewing distance, so the latest viewing distance has become 5.5H, much smaller than Tanton's number [17].

When display technologies are discussed and compared, people tend to compare by the catalog specification, such as viewing angle. However, the survey of viewing condition shows that none watches television from 170° angle direction. To make real high-performance display happen, it is important to consider the range of viewer's location and maximize the display performance, such as contrast ratio, for it.

4.3 PASSIVE-MATRIX OLED DISPLAY

Actual display design methodologies are discussed in the following chapters. There are several forms of OLED displays, such as segment displays (involving formation of electrodes by photolithography to create a pattern where the emission needs to take place and no matrix is formed), passive-matrix displays [18–22], and active-matrix displays. The passive-matrix display design is discussed here.

4.3.1 Structure

Figures 4.20 and 4.21 show passive-matrix OLED (PMOLED) displays (also called *passive OLED displays*). The data line, which is oriented vertically, and scan line, which is oriented horizontally, are formed, and the OLED devices are fabricated between the lines. Red, green, and blue subpixels are allocated to specific lines; data lines are driven by data drivers, and scan lines are driven by scan drives. (Data line and scan line can also be driven by a display controller chip to reduce cost in actual case.)

Figure 4.22 shows a microscopic view of a passive-matrix OLED display. When voltage is applied to data and scan lines, current flows in the target pixel and emission takes place.

The most common OLED device patterning method involves the shadow masking process (discussed in Section 5.1.1). For cathode patterning, an overhanging photoresist (cathode separator) can be applied with cathode self-alignment as shown in Figure 4.23. The pixel definition layer (PDL; also called "edge cover" or "edge insulator") is also described in Figure 4.23. It is used (1) to avoid defective pinhole formation made by the steep anode edge, which sometimes causes an electric short between anode and cathode and (2) to apply accurate pixel area definition by photo patterning, which can facilitate more accurate positioning of the patterning edge than shadow masking. The PDL is normally made of photoresist.

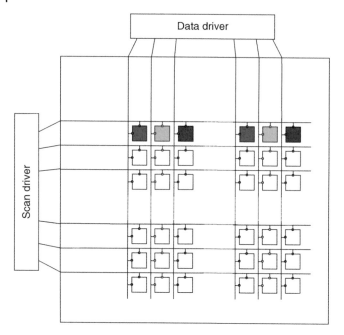

Figure 4.20 Matrix driving of a passive-type OLED display.

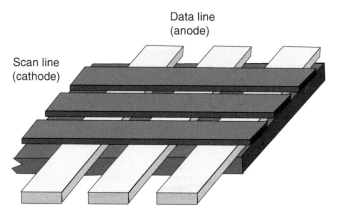

Figure 4.21 Structure of a passive-matrix OLED display.

4.3.2 Pixel Driving

The equivalent circuit of an OLED device is depicted in Figure 4.24, with a diode having parasitic capacitance. When the organic layer thickness is ~100–200 nm, and if the pixel area is 0.2 mm^2, the capacitance is somewhere in picofarads. The capacitance charge/discharge current does not contribute

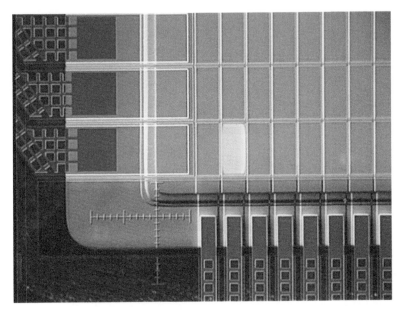

Figure 4.22 Microscopic view of a passive-matrix OLED display.

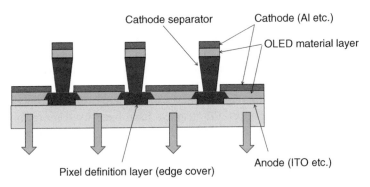

Figure 4.23 A passive-matrix display employing the cathode separator patterning method.

to the OLED emission, so it can increase the power consumption and reduce the emission period. Therefore, it is very important to minimize the parasitic capacitive current when designing passive-matrix OLED displays.

When driving a pixel as shown in Figure 4.22, it is possible to transfer charge (precharging) from the most recently addressed pixels to the next row of pixels by applying a voltage lower than the threshold voltage to "off" lines. In this way, crosstalk is avoided and the charging period is reduced by avoiding unnecessary charging.

Data line

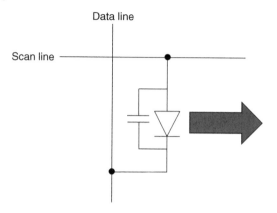

Figure 4.24 Pixel circuit of a
passive-matrix OLED display.

Scan line

Also, crosstalk (luminance error due to unintended signals) takes place according to the precharging situation difference. To avoid this phenomenon, the data line and the scan line are connected together electronically, at the onset of every frame period, to prepare for emission in order to reset the precharging levels.

With simple scanning of a passive-matrix OLED display, emission of OLED devices takes place only for one selected line. For example, if the number of the scan line is N, the emission period in one timeframe $(t_{timeframe})$ is $t_{timeframe}/N$, and is very short. Normally, the OLED device lifetime is proportional to $1/L^n (n = 1.2-1.9$, where L is luminance (see discussion in Section 2.5), so higher resolution can shorten the OLED device lifetime. To correct this problem, multiline scanning, which drives multiple lines at the same time, is used (Figure 4.25).

Figures 4.26 and 4.27 show examples of actual passive-matrix OLED display products. Thus far, passive-matrix OLED displays have been used more for lower-resolution displays than high-resolution displays because of the following:

- High luminance (needed for high resolution) required due to short emission time, which diminishes lifetime
- Insufficient charging period for high-resolution displays, due to the time loss caused by the charging of the huge parasitic capacitance of passive-matrix OLED displays
- Insufficient data line capacitance discharging in a short time, which makes inaccurate low grayscale presentation.

Currently, passive-matrix OLED displays are used mainly for the applications that do not require high resolution, such as car audio systems and MP3 players.

Figure 4.25 A passive-matrix OLED display prototype employing a multiline scanning method (2.2-in.-scale demonstration by TDK at CEATEC 2009).

4.4 ACTIVE-MATRIX OLED DISPLAY

As discussed in the preceding section, active-matrix driving is necessary when larger or higher-resolution displays are needed. Active-matrix OLED display design techniques are discussed in this section.

4.4.1 OLED Module Components

Figure 4.28 shows an example of an active-matrix OLED (AMOLED) product. Active-matrix OLED display includes the following components:

- TFT (thin-film transistor) substrates (also called *backplane substrates*; discussed in Chapter 6)
- Encapsulation glass (discussed in Section 3.3)
- Driver integrated circuits (IC) (discussed in Sections 6.4.4 and 6.4.5)
- Circular polarizers (may be unnecessary in certain applications).

As TFT is manufactured in large scale due to LCD needs, its cost is reduced enough compared with other components. Various TFT technologies have been tried for OLED driving, such as laser-crystallized polysilicon TFT [23],

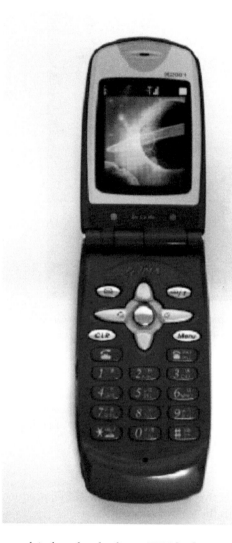

Figure 4.26 A passive-matrix OLED display product (FOMA N2001).

metal-induced polysilicon TFT [24], amorphous silicon TFT [25], microcrystalline silicon TFT [26], oxide TFT [27], and organic TFTs [28]. For the mobile display applications, laser-crystallized polysilicon technology is dominant due to several merits, such as reduced number of electrical connection, narrow boarder, peripheral circuit cost reduction, and low power operation, brought by the CMOS circuit integration capability of polysilicon TFTs. For the large television, oxide TFT is dominant because it can be manufactured with almost the same mature equipment infrastructure of amorphous silicon TFTs, while equipping with much better mobility and stability than amorphous silicon.

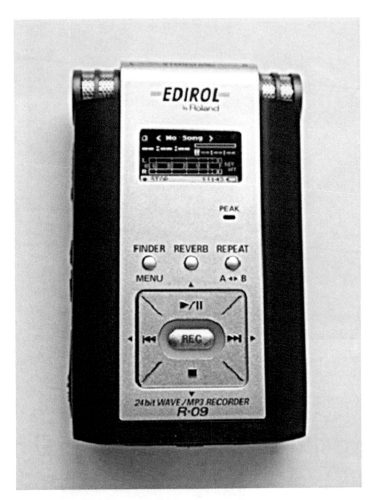

Figure 4.27 A white monochrome passive-matrix OLED display product (EDIROL R09 digital audiorecorder).

The details are discussed in Chapter 6.

For the reduction, the main focus for active-matrix should be encapsulation, driver ICs, and material utilization.

4.4.2 Two-Transistor One-Capacitor (2T1C) Driving Circuit

Figure 4.29 shows an example of the OLED pixel emission of an actual product. To control the luminance of an OLED device, TFT needs to be serially connected to OLED. There are two types of connection of TFT and OLED as shown in Figure 4.30, depending on placing TFT in (a) higher-voltage side like

Figure 4.28 A smart phone with round-shape bezel by flexible OLED display (in SID Display Week 2015 exhibition by LG display).

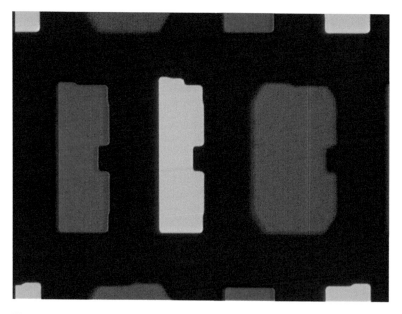

Figure 4.29 An example of an OLED pixel used in an actual product.

Figure 4.30 Two ways of OLED-TFT connection.

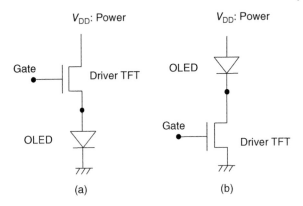

V_{DD}: Power V_{DD}: Power

Gate   Driver TFT   OLED

OLED   Gate   Driver TFT

(a)   (b)

Figure 4.31 Expression of NMOS FET and PMOS FET.

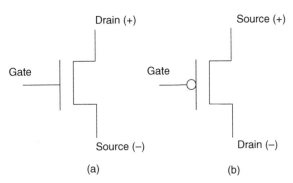

Drain (+)   Source (+)

Gate   Gate

Source (−)   Drain (−)

(a)   (b)

or (b) lower-voltage side like. As commonly used, NMOS FET and PMOS FET are described as shown in (a) and (b) of Figure 4.31, respectively.

In most of the displays, scan lines are selected and then deselected one after another. For example, 720 progressive (720p) display format driven at 60 Hz, 720 lines are selected and deselected one by one in 1/60 s. Therefore, the selected period of one scan line is $1/720/60 \cong 20\,\mu s$. If the OLED-TFT connection such as active-matrix LCDs is simply applied to the display matrix as shown in Figure 4.32, OLED in a pixel emits light for only 20 μs in one frame = 1/60 s in the case of 720p, which cannot make high luminance display performance happen. The paper by Brody et al. in 1975 employed two TFTs and one capacitor (called 2T1C circuit) as shown in Figure 4.33 [29]. With this pixel circuit, the signal data is stored in pixel capacitor during selected period and the stored voltage drives the driver TFT for a frame period, which dramatically improve the emission time that promises the high luminance OLED display operation. This 2T1C circuit is still used as the basis of most pixel circuit of OLED displays.

As mentioned above, this acronym is derived from the fact that the circuit contains two transistors and one capacitor. Other pixel circuits are equipped

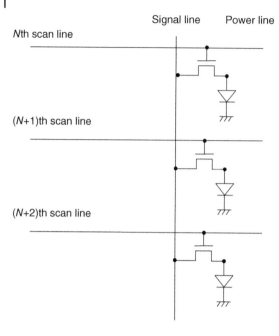

Signal line   Power line

Nth scan line

(N+1)th scan line

(N+2)th scan line

Figure 4.32 Active-matrix driving by one TFT per pixel approach.

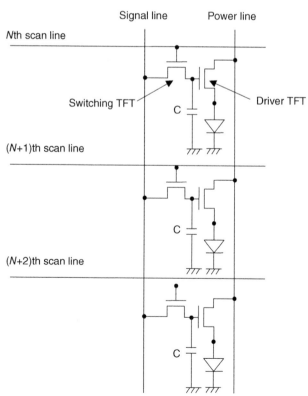

Signal line   Power line

Nth scan line

Switching TFT

Driver TFT

(N+1)th scan line

(N+2)th scan line

Figure 4.33 Active-matrix driving by two TFT and one capacitor (2T1C) per pixel approach.

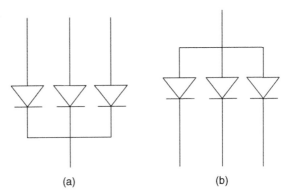

Figure 4.34 Common-cathode and common-anode connection.

(a)   (b)

with two TFTs and one capacitor; however, the designation 2T1C normally indicates a circuit configuration as shown in Figure 4.33.

Figure 4.34 shows the so-called "Common-cathode" and "Common-anode" configuration, which is frequently used in the electronic circuit. In a display using 2T1C pixels, the circuit can be formed by either of these two connections. The driver TFT, which drives the OLED current, can be either NMOS or PMOS type, so there are four kinds of 2T1C circuit, as shown in Figures 4.35 and 4.36. In the case of "NMOS + common cathode" (Figure 4.35a) and "PMOS + common anode" (Figure 4.36a), gate-to-source voltage V_{GS}, which determines the OLED current, is subjected to the OLED driving voltage variation. This configuration is called source-follower type. On the other hand, the case of "NMOS + common anode" (Figure 4.35b) and "PMOS + common cathode" (Figure 4.36b) V_{GS} is kept the same regardless of OLED voltage variation, which assures the constant OLED current. These four configurations

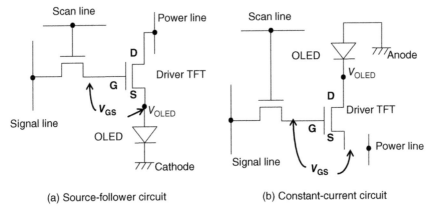

(a) Source-follower circuit   (b) Constant-current circuit

Figure 4.35 Two ways of pixel configuration using NMOS driver TFT (omitting capacitor to avoid complexity).

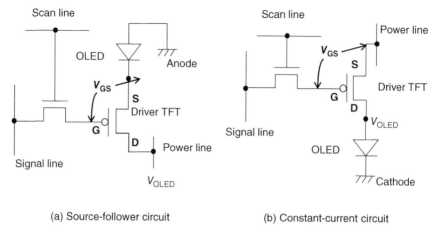

(a) Source-follower circuit   (b) Constant-current circuit

Figure 4.36 Two ways of pixel configuration using PMOS driver TFT (omitting capacitor to avoid complexity).

have a possibility to be used for either top emission or bottom emission with RGB-pixelation or with White + Color Filter method. It is important to carefully consider the pros and cons of each technology combination when a new OLED display is designed.

Currently, most of the OLED devices on the market is using anode as the bottom electrode and cathode as top electrode. (It is known that the top anode structure gives worse performance compared to top cathode structure. There are several papers reporting the performance improvement of top anode structure [30, 31].) Naturally, it makes the situation simpler if the anode is connected to TFT electrodes. In that sense, constant-current type "PMOS + common cathode" device (Figures 4.36b and 6.13) and source-follower type "NMOS + common cathode" (Figures 4.35a and 4.37) are more commonly used. Former is an industry-standard technology for mobile displays and the latter is also an industry standard for large OLED television using oxide semiconductors (Section 6.5.5).

As discussed, in the case of source-follower-type pixel configuration, the OLED current is altered by an OLED voltage increase, which happens after driving stress. To avoid the source-follower-type configuration using NMOS TFT, TFT structure as shown in Figure 4.38 is necessary. In Figure 4.38, TFT electrode is connected to the cathode of NMOS TFT via hole and overhanging photoresist structure. In this implementation, the anode is made of aluminum and forming as a low resistive line. (Anode line structure.)

Though the V_{GS} fluctuation of source-follower-type is sometimes regarded as disadvantageous, there are actually some merits using it. In the case of constant-current configurations regardless of NMOS or PMOS, the source electrode is connected to the power line, as shown in Figures 4.35b and 4.36b.

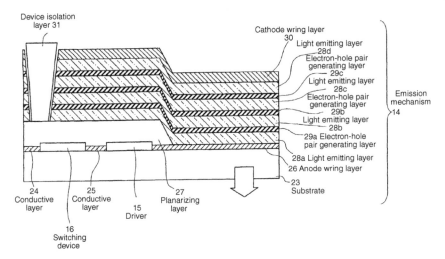

Figure 4.37 An example of common-cathode tandem white OLED pixel configuration by NMOS TFT [32].

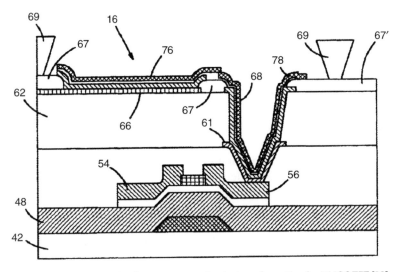

Figure 4.38 An example of common-anode pixel configuration by NMOS TFT [33].

Due to this reason, the OLED current is subjected to the voltage drop of the power line. However, in the case of source-follower configuration, the source node is not connected to the power line, which decreases the sensitivity to the power line voltage drop.

There is another merit of source-follower type. In the case of constant-current circuit, when $I–V$ curve of OLED device is fluctuated, the OLED current would

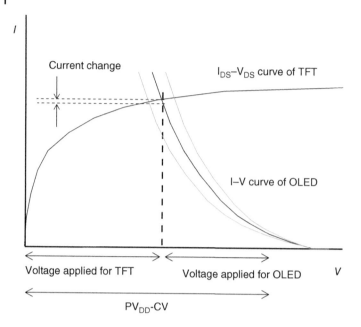

Figure 4.39 OLED current change due to OLED performance change in source-follower type.

be largely changed. However, in the case of source-follower-type configuration, unique compensation happens. As shown in Figure 4.39, because the TFT current is saturated, the OLED current does not change very much when OLED characteristics are changed. Similarly, when TFT characteristics are changed, OLED current change is reduced as shown in Figure 4.40.

In any case, OLED voltage rise, due to the driving stress has been dramatically reduced these days for the product-level OLED device, so the risk of using source-follower type is a minor issue already.

For actual implementation, more complicated circuits are often used to compensate for the current distribution of the driver TFT and OLED device or for their instabilities. These techniques are discussed in Section 6.4.4.

Figure 4.41 shows the matrix arrangement of pixels. Color can be expressed by the combination of red, green, and blue subpixels. Most data drivers can handle 6- or 8-bit grayscales, and for special applications, such as medical use, 10 or 12 bits are used.

When an 8-bit grayscale is used, each (R, B, or G) subpixel can display $2^8 = 256$ grayscales (an average computer handles the value from 0 to 255:256 grayscale in total); therefore, $256^3 = 16,777,216$.

Almost 16 million colors can be expressed; this is sometimes called a *full-color display*.

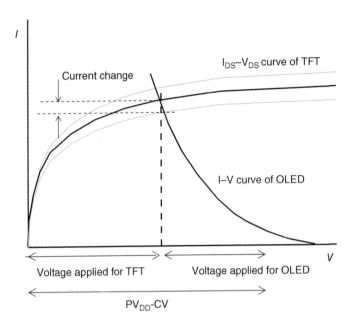

Figure 4.40 OLED current change due to TFT performance change in source-follower type.

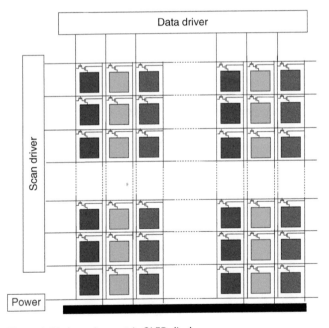

Figure 4.41 An active-matrix OLED display.

As an OLED display is self-emissive, a high contrast value can be obtained if measured in a dark room. However, displays are rarely used in dark environments, so it is more useful to measure contrast ratio in actual environments. For mobile applications, legibility of characters and visibility in bright ambient light are required, so the surface treatment of a display is very important.

4.4.3.1   Living Room Contrast Ratio

As discussed, a dark contrast ratio value does not always indicate the quality of a display. Self-emissive displays show good performance in darker environments, but the image quality is degraded in brighter environments. Table 4.3 lists illuminance values in various environments.

For television applications, the contrast ratio found in home living rooms, which represents nearly the darkest illuminance in actual usage, is often discussed. This contrast ratio value in a 200 lx environment is sometimes called the *living room contrast ratio* (see Section 4.2.1.4 for detailed equation).

Figure 4.42 shows the living room contrast ratio simulation results of a $500 \, cd/m^2$ display, with various surface reflection values without a circular polarizer; however, surface reflection can be more than 1% without it because of reflection due to electrode and wiring. Color filters are also used to absorb the reflection to enhance the living room contrast ratio of an OLED display.

On the other hand, state-of-the-art premium LCD TV displays use expensive antireflection to obtain a ~1500–1800 : 1 living room contrast ratio,

Table 4.3 Illuminance in Various Environments

Location/Environment	Illuminance (lx)
Sunny place in fine spot on warm summer day	>100,000
Shade in hot weather	10,000
Office	750
Reception room in an office	500
Conference room	300–700
Classroom	500
Showroom	500–1500
Display shelf in a shopping mall	500
Living room	200–300
Hotel room	100
Museum gallery	50
Cinema	3
Night with full moon	0.2

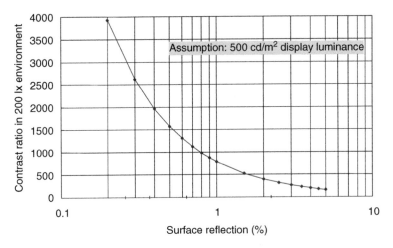

Figure 4.42 Plot of contrast ratio versus surface reflection in a living room.

so an OLED TV display may need to include a circular polarizer with high-performance antireflection.

4.4.3.2 Chroma Reduction Due to Ambient Light

Many display technologies are claiming large color reproduction in their advertisement or in the brochures these days. However, it should be noted that surface reflectance significantly reduces the color reproduction of a display, so catalog specification is not always true in ambient environment.

For example, if a display is used in illuminance E and if surface reflectance is R, luminance increase caused by reflected ambient light can be calculated as follows:

$$\Delta L = \frac{E \cdot R}{\pi} \tag{4.29}$$

For example, if $E = 10,000$ lx (shade in bright sunshine, Table 4.3), $R = 1\%$, luminance of reflected light would be

$$\Delta L = \frac{10000 \cdot 0.01}{\pi} = 31.8 \ [\mathrm{cd/m^2}]$$

It is as if all the pixels in a display get additional 31 cd/m² white (environmental color) emission. Even when pure colors, such as red, green, and blue, are displayed, 31 cd/m² white is always added to the displayed colors, which generates pink, light green, and sky blue, respectively. This effect reduces the color reproduction of a display, so low surface reflection is quite important for high-performance display.

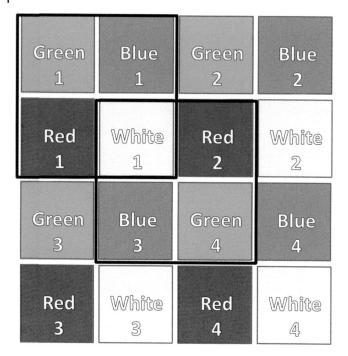

Figure 4.43 An example of an RGBW (RGB+white) pixel layout.

4.4.4 Subpixel Rendering

Figure 4.43 shows four sets of an RGBW subpixel (red, green, blue, and white) layout. As discussed in Section 7.4, the additional white subpixel provides the advantage of power reduction. The primary colors red, green, blue, and white can be used, for example, to show one pixel color (green—1, blue—1, red—1, white—1); therefore, Figure 4.43 potentially shows four pixels.

However, there are other ways of driving this subpixel set. A close scrutiny of Figure 4.43 reveals that the (white—1, red—2, blue—3, green—4) subpixel set can also be regarded as a pixel (dashed line). This kind of approach to increase the apparent resolution is called the *subpixel rendering* method, and the apparent resolution obtained by this method is sometimes called *pseudoresolution*. There are various kinds of subpixel rendering methods, such as the following:

1. Pixel configuration using red, green, blue, and white [34–36]
2. Pixel configuration by combination of large blue + green and large red + green, or two blue, one red, and one green subpixels; this is known as the *PenTile*® (Samsung Electronics) *configuration*. PenTile also includes some implementations using the RGBW pixel layout [37, 38].

The subpixel rendering method is an effective approach to increase the apparent display resolution with only minimal increase in IC cost.

References

1 M. O'Regan, D. Lecloux, C. Hsu, E. Smith, A. Goenaga, and C. Lang, Solution processing of small molecule OLED materials at DuPont displays, *IMID/IDMC '06 Digest*, p. 1689 (2006).

2 "IHS: A 5" FHD AMOLED costs less than an equivalent LCD", OLED-Info, http://www.oled-info.com/ihs-5-fhd-amoled-costs-less-equivalent-lcd.

3 "UBI Research: LG to continue OLED TV cost reduction, will reach a premium of 10% over LCD", OLED-Info, http://www.oled-info.com/ubi-research-lg-continue-oled-tv-cost-reduction-will-reach-premium-10-over-lcd.

4 L. Ronchi, The Purkinje effect: some considerations on the interplay of receptor postreceptoral mechanisms, *Color Res. Appl.* **16**(1): 10-15 (1991).

5 R. Donofrio, Review paper: the Helmholtz–Kohlrausch effect, *J. SID* **19**(10):658–664 (2011).

6 J. Sanchez and M. Fairchild, Quantification of the Helmholtz–Kohlrausch Effect for CRT Color Monitors, 9th Congress of the International Colour Association, *Proceedings of SPIE*, Vol. 4421 (2002).

7 T. Tsujimura, W. Zhu, S. Mizukoshi, N. Mori, K. Miwa, S. Ono, Y. Maekawa, K. Kawabe, and M. Kohno, Advancements and outlook of high performance active-matrix OLED displays, *SID 2007 Digest*, pp. 84–88 (2007).

8 C. J. Bartleson and E. J. Breneman, Brightness perception in complex fields, *J. Opt. Soc. Am.* **57**(7):953–957 (1967).

9 M. Hack, M. Weaver, J. Brown, L. Chang, C. Wu, and Y. Lin, AMLCD and AMOLEDs: how do they compare for green energy efficiency?, *SID 2010 Digest*, Vol.41, Issue 1. pp. 894–897 (2010).

10 M. R. Pointer, The Gamut of real surface colours, *Color Res. Appl.* **5**(3):145–155 (2007).

11 D. L. McAdam, Visual sensitivities to color differences in daylight, *J. Opt. Soc. Am.* **32**(5):247–274 (1942).

12 J. Someya, Y. Inoue, H. Yoshii, M. Kuwata, S. Kagawa, T. Sasagawa, A. Michimori, H. Kaneko, and H. Sugiura, Laser TV: ultra-wide gamut for a new extended color-space standard, xvYCC, *SID 2006 Digest*, pp. 1134–1137 (2006).

13 Y. Hisatake, A. Ikeda, H. Ito, and M. Obi, Yasushi Kawata and Akio Murayama, *SID Symposium Digest of Technical Papers*, Vol. **39**, Issue 1, pp. 1316–1319 (2008).

14 S. Kubota, J. Shimada, S. Okada, Y. Nakamura, and E. Kido, Television viewing conditions at home, *J. ITE* **60**(4):597–603 (2006).

15 K. Moriguchi and R. Yoshitake, Acceptable range of observation angles for moving images, *The 15th International Ergonomics Association Congress* (2003).

16 N. Tanton, Results of a Survey on Television Viewing Distance, BBC R&D White Paper, WHP090 (2004).

17 K. Noland and L. Truong, A Survey of UK Television Viewing Conditions, BBC R&D White Paper, WHP287 (2015).

18 Y. Fukuda, S. Miyaguchi, S. Ishizuka, T. Wakimoto, J. Funaki, H. Kubota, T. Watanabe, H. Ochi, T. Sakamoto, M. Tsuchida, I. Ohshita, H. Nakada, and T. Tohma, Organic LED full color passive-matrix display, *SID 1999 Digest*, p. 430 (1999).

19 Y. Chang, M.-K. Wei, C.-M. Kuo, S.-J. Shieh, J.-H. Lee, and C.-C. Chen, Manufacturing of passive matrix OLED—organic light emitting display, *SID 2001 Digest*, p. 1040 (2001).

20 Y. Sakaguchi, H. Tada, T. Tanaka, E. Kitazume, K. Mori, S. Kawashima, and J. Suzuki, Color passive-matrix organic LED display using three emitters, *SID 2002 Digest*, p. 1182 (2002).

21 C. MacPherson, M. Anzlowar, J. Innocenzo, D. Kolosov, W. Lehr, M. O'Regan, P. Sant, M. Stainer, S. Sysavat, and S. Venkatesh, Development of full color passive PLED displays by inkjet printing, *SID 2003 Digest*, p. 1191 (2003).

22 M. Fleuster, M. Klein, P. V. Roosmalen, A. D. Wit, and H. Schwab, Mass manufacturing of full color passive-matrix and active-matrix PLED displays, *SID 2004 Digest*, p. 1276 (2004).

23 M. Kimura, I. Yudasaka, S. Kanbe, H. Kobayashi, H. Kiguchi, S. Seki, S. Miyashita, T. Shimoda, T. Ozawa, K. Kitawada, T. Nakazawa, W. Miyazawa, H. Ohshima, Low-temperature polysilicon thin-film transistor driving with integrated driver for high-resolution light emitting polymer display, *IEEE Trans. Electron Dev.* **46**(12) (2002).

24 S. Y. Yoon, K. H. Kim, C. O. Kim, J. Y. Oh, and J. Jang, Low temperature metal induced crystallization of amorphous silicon using Ni solution, *J. Appl. Phys.* **82**(2):5865–5867 (1997).

25 T. Tsujimura et al., A 20-inch OLED display driven by super-amorphous-silicon technology, *SID 2003 Proc.*, p. 6 (2003).

26 T. Tsujimura, Amorphous/microcrystalline silicon thin film transistor characteristics for large size OLED television driving, *Jpn. J. Appl. Phys.* **43**(8A):5122–5128 (2004).

27 K. Nomura, A. Takagi, H. Ohta, T. Kamiya, M. Hirano, and H. Hosono, Amorphous oxide semiconductors for high-performance flexible thin-film transistors, *Jpn. J. Appl. Phys.* .**45**:4303–4308 (2006).

28 Y. Inoue, Y. Fujisaki, T. Suzuki, S. Tokito, T. Kurita, M. Mizukami, N. Hirohata, T. Tada, and S. Yagyu, Active-matrix OLED panel driven by organic TFTs, *International Display Workshop 2004*, pp. 355 (2004).

29 T. P. Brody, F. C. Luo, Z. P. Szepesi, and D. H. Davies, A 6 × 6-in 20-lpi electroluminescent display panel, *IEEE Trans. Electron Dev.* **22**(9):739–748 (1975).

30 J. Noh, M. Kang, J. Kim, J. Lee, Y. Ham, J. Kim, and S. Son, Inverted OLED, *SID Symposium Digest of Technical Papers*, Vol. 39, Issue 1, pp. 212–214 (2012).

31 H. Hosono, J. Kim, T. Kamiya, N. Nakamura, and S. Watanabe, Novel inorganic electron injection and transport materials enabling large-sized inverted OLEDs driven by oxide TFTs, *SID Symposium Digest of Technical Papers*, Vol. 47, Issue 1, pp. 401–404 (2016).

32 T. Tsujimura et al. US7348944 (2008).

33 T. Tsujimura, Organic led device, US6727645 (2004).

34 M. Haruhiro, T. Ueki, Y. Oana, and M. Kajimura, Color Liquid Crystal Display Device and Manufacture thereof, Jpn. Patent 1,059,318 (1989).

35 C. H. Brown Elliott and T. L. Credelle, PenTile matrix displays and drivers, *ADEAC 2005 Digest*, p. 87 (2005).

36 T. L. Credelle and C. H. Brown Elliott, High-pixel-density PenTile matrix RGBW displays for mobile applications, *IMID 2005 Digest*, p. 867 (2005).

37 T. L. Credelle, C. H. Brown Elliott, and M. F. Higgins, MTF of high-resolution PenTile matrix displays, *Eurodisplay 2002 Digest*, p. 159 (2002).

38 C. H. Brown Elliott and M. F. Higgins, New pixel layout for PenTile matrix, *IDMC 2002 Digest* (2002).

5

OLED Color Patterning Technologies

To ensure the success of OLED technology as a major industry, it is necessary
to develop strategies to achieve low cost and high productivity, and to develop
improved displays to compete with conventional displays and be introduced in
new applications on the market. In this chapter, we discuss the color patterning
technologies, which gives major contribution to cost and productivity.

5.1 COLOR-PATTERNING TECHNOLOGIES

As OLED materials degrade under at high temperature and humidity, it is very
difficult to pattern the organic layer by means of photo-patterning technology,
which is normally used in semiconductors and TFTs. Therefore, the shadow
mask technique has been used for the patterning, but this leads to many problems. Various approaches have been developed to deal with these problems.

5.1.1 Shadow Mask Patterning

5.1.1.1 Shadow Mask Process

Currently (as of 2017), most small- to medium-size OLED display products
for mobile application use shadow masks for color patterning, as shown in
Figures 5.1 and 5.2.

Criteria for optimal patterning include the following:

1. High-dimensional accuracy
2. High aperture accuracy
3. Low thermal expansion
4. Minimal shadowing effect (unintended excessive shadowing due to
   high-incident-angle incident molecules relative to the normal angle, which
   are prevented from reaching the substrate).

These criteria are important in maintaining display product quality. If they are
not satisfied, an abnormal image such as that shown in Figure 5.3 is displayed

OLED Display Fundamentals and Applications, Second Edition. Takatoshi Tsujimura.
© 2017 John Wiley & Sons, Inc. Published 2017 by John Wiley & Sons, Inc.

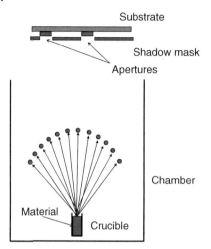

Substrate

Shadow mask

Apertures

Chamber

Material

Crucible

Figure 5.1 Patterning using a shadow mask.

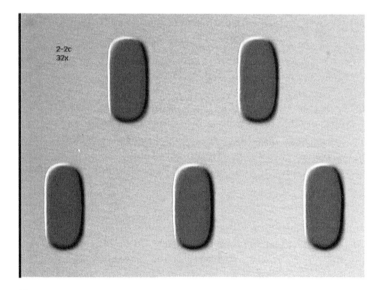

2-2c
32x

Figure 5.2 Microscopic view of a shadow mask.

and causes yield degradation, so it is very important to take these criteria into consideration.

To reduce the shadowing effect, use of a thin shadow mask with a tapered shape (Figure 5.4) in its aperture provides an effective method.

A *shadow mask* is a thin film of metal or an alloy such as invar metal (Ni alloy), which is normally made by an *etching* method or by electroplating. Features of the *etching* method are as follows:

Figure 5.3 Color variation
due to shadow mask
deformation.

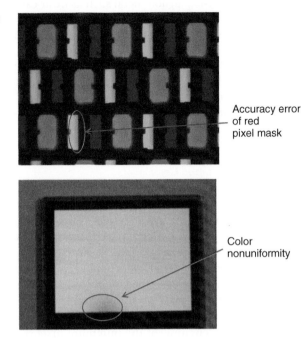

Accuracy error
of red
pixel mask

Color
nonuniformity

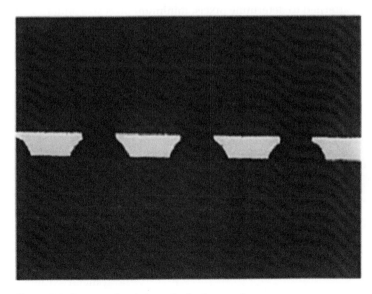

Figure 5.4 Cross-sectional view of a shadow mask.

- A low-thermal-expansion metal such as invar can be used.
- It is relatively inexpensive.
- The corner of the shadow mask pixel area is rounded.

In contrast, the *electroplating* method can achieve a very thin mask, so it is better than etching in terms of the shadowing effect; however, thermal expansion of most electroplating type mask is not as low as with the etching method, and electroplating is normally more costly. So, the etching and electroplating methods both have pros and cons.

5.1.1.2 Blue Common Layer

Figure 5.5 shows the Blue Common Layer (BCL) structure, reported by Kim et al. [1]. In the case of red and green device, the recombination zone is located at HTL/EML interface, which results in no blue emission, as the holes are blocked by the HOMO energy gap between HTL and EML (Figure 5.6). In the case of blue device, as no EML layer exists, holes are accumulated at the HTL/BCL interface, which results in blue emission (Figure 5.7).

The alignment error described in Section 5.1.1.1 and various design margins define the minimum pixel size limitation. However, there is a trend to aim at higher resolutions to gain better user experience.

BCL idea is to circumvent the minimum pixel size limitation by using blue emission layer as common layer of green and red pixels. As blue layer does not have to be patterned to determine pixels, minimum pixel size limitation is relaxed by the factor of 1/3.

The idea is also used for solution-processed device fabrication [2].

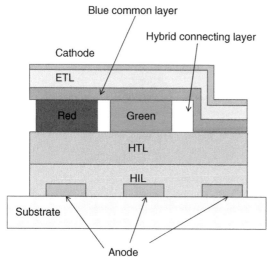

Figure 5.5 Blue Common Layer structure.

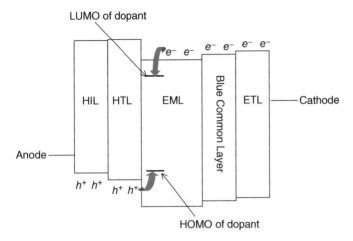

Figure 5.6 Energy diagram of red device in Blue Common Layer Structure.

Figure 5.7 Energy diagram of blue device in Blue Common Layer Structure.

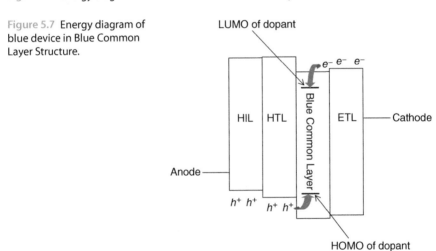

5.1.1.3 Polychromatic Pixel

Figure 5.8b shows the polychromatic pixel, fabricated by vertically stacked OLED unit [3]. (The process details are discussed in Section 9.5.) In conventional pixel design, shown in Figure 5.8a, red, green, and blue subpixels are arranged next to each other so that the mixed color can be observed when the viewing distance is far enough compared with the pixel pitch. With this arrangement, high pixel current density is required to have the same display luminance due to poor aperture ratio, compared with the polychromatic pixel arrangement. High current density causes significant lifetime loss as discussed in Section 2.5.2. Contrally, polychromatic pixel arrangement has very wide

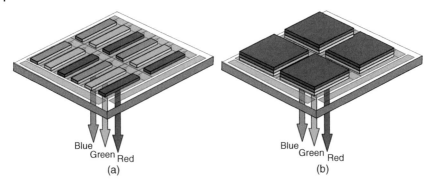

Blue
Green Red
(a)

Blue Green Red
(b)

Figure 5.8 Conventional pixel arrangement (a) and polychromatic pixel arrangement (b).

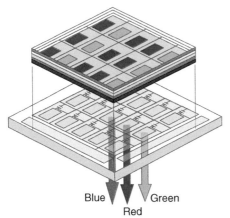

Blue Green
Red

Figure 5.9 Schematic depiction of the RGB pixelation method.

emission area in a pixel, so the low pixel current density reduces the pixel luminance decay, which will lead to much longer image sticking lifetime.

5.1.2 White + Color Filter Patterning

Figure 5.9 illustrates the conventional RGB pixelation method (also called the *RGB side-by-side method*). Red, green, and blue emissive pixels are patterned by, for example, shadow masking. On the other hand, Figure 5.10 illustrates the *white + color filter method*, which displays color by white OLED device emission [4] and color filter absorption of emitted light to tune the color.

The shadow masking process, which patterns the OLED device, sometimes causes production yield issues such as shadow mask deformation and particle generation by shadow mask–TFT substrate contact. The white + color filter method has an advantage in terms of high production yield; however, it is necessary to use some technologies to reverse the emission efficiency loss due to color filter absorption. Details are discussed in Chapter 7.

Figure 5.10 Schematic representation of the white + color filter method.

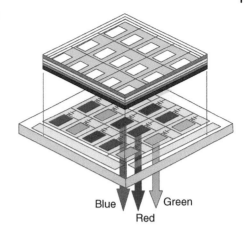

Blue   Green
Red

5.1.3   Color Conversion Medium (CCM) Patterning

Filtering methods other than the white + color filter method are also used to display color images. The *color conversion medium* (CCM) method [5] consists of the conversion of blue emission to a longer wavelength using a CCM layer. Unlike the color filter method, which absorbs unnecessary wavelengths and generates heat, the CCM method downconverts the light, so this method can be effective at least theoretically. However, the current CCM method entails application of a color filter to ensure good color quality and also requires a high-performance (high-efficiency and long-lifetime) blue emitter, the performance of which is normally the most difficult to maintain among all the colors, so the CCM method is not widely used these days.

5.1.4   Laser-Induced Thermal Imaging (LITI) Method

The laser-induced thermal imaging (LITI) method uses laser irradiation onto a light-to-heat conversion (LTHC) layer. The resulting local temperature rise transfers organic materials onto the substrate in a predetermined pattern. (The RIST method, discussed in Section 5.1.5, also uses laser irradiation but involves reevaporation, not material transfer as does LTHC.) Figure 5.11 shows the process flow of the LITI method.

The LTHC layer is formed on the donor film. The organic material can be either polymer OLED or small-molecule OLED. By means of laser irradiation, the LTHC layer absorbs light and generates heat. In response to the heat, the organic material layer detaches from the donor film and is transferred to the substrate.

Thus, an organic material pattern is formed on the substrate at the target location. This method can transfer either an organic material layer (donor film contains only one layer in this case) or multiple layers simultaneously (multiple layers are deposited on the donor film in advance in this case).

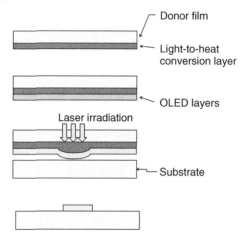

Donor film

Light-to-heat
conversion layer

OLED layers

Laser irradiation

Substrate

Figure 5.11 Schematic representation of organic material patterning using the laser-induced thermal imaging (LITI) method.

Figure 5.12 An active-matrix OLED display obtained using the LITI method (14-in. demonstration at SID 2004 by Samsung SDI).

Lee et al. [6] report that the donor and receptor surfaces must be in intimate contact for this process to happen. The donor film can be prepared by spin coating, web coating, or vacuum evaporation. The irradiation system used in the paper was continuous-wave 8.0 W Nd : YAG laser.

Figure 5.12 shows an example of an active-matrix OLED display prototype fabricated using the LITI method.

5.1.5 Radiation-Induced Sublimation Transfer (RIST) Method

The radiation-induced sublimation transfer (RIST) (also called *laser-induced patternwise sublimation* [LIPS]) method involves reevaporation of organic material by laser irradiation.

Figure 5.13 shows the process flow of the RIST method. Initially, organic materials are deposited by thermal evaporation on the donor substrate. The donor substrate can be either glass or plastic. Following irradiation of laser light to the organic film on the donor substrate, reevaporation occurs locally. With this approach, organic film is deposited on the substrate at the target location from the donor substrate.

Figure 5.14 shows an example of an active-matrix OLED display prototype created using the RIST (LIPS) method.

Boroson et al. [7] report that the transfer process requires vacuum environment with the donor–receptor transfer gap of 1–10 μm. 810 nm diode laser and flash lamps with optical mask are used for the irradiation. The donor includes polyimide support (75 μm), a silicon antireflection layer (50 nm), a chromium absorption layer (40 nm), and an organic transfer layer (20–60 nm). The organic transfer layers are typically evaporated in vacuum.

As discussed, RIST uses laser to transfer the organic pattern from donor glass to device substrate, but there is another method proposed to circumvent the disadvantages of lasers, such as maintenance cost. Flash Mask Transfer Lithography (FMTL) [8] utilizes flash lamp to do the transfer instead of laser equipment.

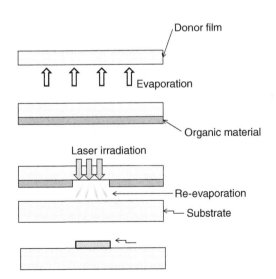

Figure 5.13 Schematic representation of organic material patterning using the radiation-induced sublimation transfer (RIST) method.

Figure 5.14 An active-matrix OLED display obtained using the RIST method (27-in. demonstration at CEATEC 2007 by Sony).

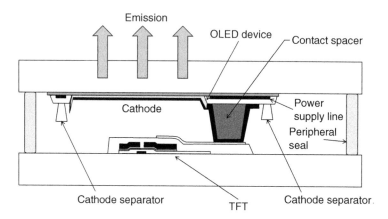

Figure 5.15 Schematic representation of a display structure obtained using the dual-plate OLED display (DOD) method.

5.1.6 Dual-Plate OLED Display (DOD) Method

The dual-plate OLED display (DOD) method fabricates TFT substrate and OLED substrate separately and combines the two substrates later. Figure 5.15 illustrates the process flow of the DOD method.

After the TFT is fabricated using a conventional TFT process, a peripheral sealing resin is dispensed on the TFT substrate. An anode bank is formed on the OLED substrate, and a conventional OLED process takes place on it. The TFT substrate is attached to the OLED substrate, and the seal resin is cured by UV irradiation to produce an OLED display (Figures 5.16 and 5.17).

If the attachment process is well engineered, the production yield of this method can be good because the best TFT and OLED substrates are selected before attachment; therefore, no yield loss happens due to the matching of bad TFT + good OLED or good TFT + bad OLED. However, it is important to secure the electrical contact by a contact spacer, as contact problem can cause yield loss, which may diminish the merit of this method.

5.1.7 Other Methods

Figure 5.18 shows the structure of a photopatternable PLED material. The PLED material has the same photosensitivity as that of normal photoresist, so it can be patterned by photoexposure. However, it is very difficult to have both good photosensitivity performance and good OLED device performance, so photopatternable PLED material is not widely used.

5.2 SOLUTION-PROCESSED MATERIALS AND TECHNOLOGIES

Though an OLED display consists of fewer components than an LCD, the cost is, in most case, still higher. The higher cost structure of OLED is mainly due to (1) high encapsulation cost, especially due to the use of cap glass, which requires expensive cavity formation, (2) low material utilization, and (3) low throughput of manufacturing equipment. To reduce (1), frit glass (Figure 3.32) and CVD encapsulation (discussed in Section 3.3.4) are used. To cope with (2) and (3),

Figure 5.16 Flowchart of the DOD process.

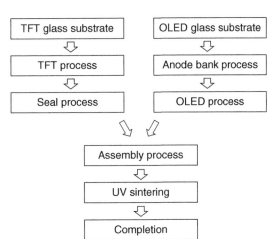

Figure 5.17 An example of an active-matrix OLED display obtained using the DOD method (19-in. demonstration at SID 2009 by LG Display).

Figure 5.18 An example of a photopatternable polymer OLED (PLED) material (Reported by Becker et al. [9].]

many methods have been proposed. An approach using solution processing is discussed in this section, and another approach using a new vacuum processing method is discussed in Section 5.3. Both cases are claiming large improvements in material utilization and throughput. Figure 5.19 shows the cost simulation of active-matrix OLED display in comparison with LCD (4-in. display driven by PMOS LTPS-TFT substrate with integrated scanner and discrete source driver IC is assumed). By applying flat encapsulation glass (such as frit technology) and improvement in (2) and (3), the cost can be reduced to almost equal to that of LCDs. (This simulation is not taking cost reduction into consideration due

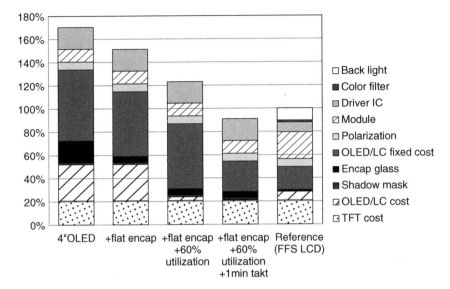

Figure 5.19 Active-matrix OLED display cost simulation.

to scale merit, so there is more room for improvement in the cost structure.) In this section, OLED device fabrication using solution processing is discussed.

To dramatically improve material utilization, several printing methods have been developed, including inkjet printing, nozzle coating [10], offset printing, and gravure printing. Here, we discuss the most popularly used method, inkjet printing.

The typical working principle of the inkjet method consists of the electrostatic acceleration and deflection of ink droplets emerging from nozzles to form designated patterns on the substrate. Ink is ejected from the nozzle by a *piezoelectric device*, which produces a mechanical force when a voltage is applied.

Application of pulse-modulated voltage to the piezoelectric device results in an instantaneous pressure increase that ejects the ink from the nozzle (Figure 5.20).

The method requires control to ensure uniform discharge speed of droplets, angle, volume, and weight. Also, to ensure optimal printing capability, the organic materials need to have the properties of a good ink, such as appropriate surface tension, viscosity, and condensation in the nozzle.

To control the placement and spreading of the ink droplets, surface energy control of the substrate by surface modification or a patterned resin (called a *bank*) is often used, as shown in Figure 5.21. With this method, accurate ink location can be achieved; this method is so reliable that a range of tens of micrometers of ink projection inaccuracy can be tolerated, as the pixel area is determined by the bank structure edge.

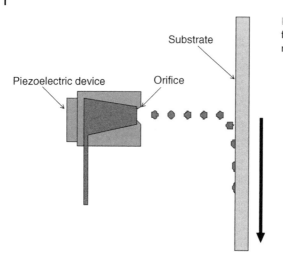

Piezoelectric device   Orifice

Substrate

Figure 5.20 Setup for thin-film formation using the inkjet method.

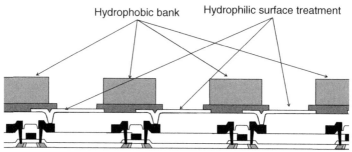

Hydrophobic bank   Hydrophilic surface treatment

Figure 5.21 Schematic representation of a hydrophilic/hydrophobic structure designed for inkjet patterning.

To enable the hydrophilic/hydrophobic patterning discussed above, surface energy control to ensure adhesion on the pixel but not on the bank is necessary. If the combined surface energy of the substrate and a droplet before and after the droplet landing is F_b and F_a, then, as the kinetic energy (E_k) must be higher than the surface energy on the bank, the following equation must be satisfied:

$$E_k > F_{a,\text{bank}} - F_{b,\text{bank}} \tag{5.1}$$

On the other hand, to ensure that the kinetic energy is lower than the surface energy in a pixel, the following criterion must be satisfied:

$$E_k < F_{a,\text{pixel}} - F_{b,\text{pixel}} \tag{5.2}$$

For this kind of control, contact angle measurement such as that depicted in Figure 5.22 is effective. For example, a hydrophilic area such as the pixel

Figure 5.22 Schematic depiction of contact angle measurement.

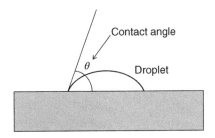

Contact angle

θ

Droplet

Figure 5.23 An OLED display obtained using the PLED inkjet method (21-in. demonstration at Display 2007 by Toshiba Matsushita Display).

requires a contact angle of $\theta < 10°$ and a hydrophobic area such as the bank requires a contact angle of $\theta > 80°$ to ensure self-aligned patterning.

Regarding the inkjet method, if the droplet is small, the surface area is large relative to its volume, and unduly rapid evaporation may result. Also, the organic materials are transferred between the central, rapid-evaporation region and the slow-evaporation, bank-side regions. Therefore, both surface tension and drying conditions must be considered.

Figure 5.23 shows a 21-in. display prototype using the polymer OLED inkjet method.

Polymer OLED (PLED) materials can contain long conjugated bonds, so red-emission OLED devices tend to perform well; however, device efficiency is lower when blue-emission OLED devices are used. This is why the use of a modifying agent, such as a dendrimer—a molecular structure with soluble chains surrounding the emission center (Figure 5.24)—has been proposed.

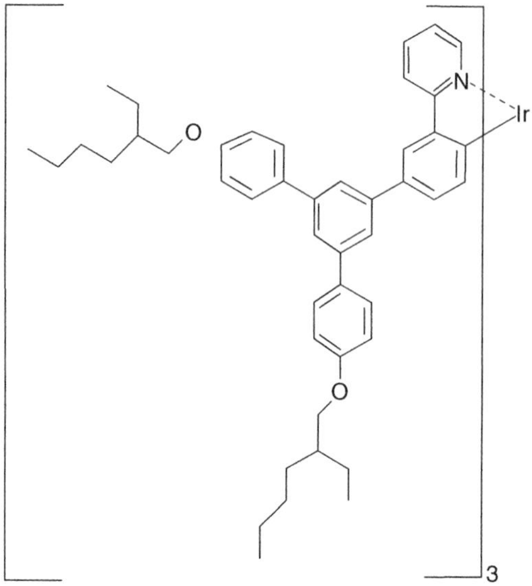

Figure 5.24 Structure of a dendrimer material (IrppyD) (Reported by Markham et al. [11]).

Dendrimers render the small-molecule material soluble and thus improve the performance and productivity of both small-molecule and PLED devices.

5.3 NEXT-GENERATION OLED MANUFACTURING TOOLS

Current OLED material is expensive, ranging from $5 to $10,000 per gram. In actual thermal evaporation manufacturing processes, OLED material deposited on the substrate constitutes only ~2–3% (including material consumption during startup, stabilization, tooling/equipment check, substrate waiting, cooldown, and unusable remained material in crucible after a week of operation) of the total amount of prepared material in the evaporation source. Low material utilization can bring high cost, so it is very critical to suppress material consumption during OLED display manufacturing. To address this situation, many high-material-utilization processes, such as the vapor injection source technology (VIST) method, hot-wall method, and organic vapor-phase deposition (OVPD), have been proposed.

5.3.1 Vapor Injection Source Technology (VIST) Deposition

The VIST method (also called Kodak vapor injection source [KVIS]) involves a procedure similar to flash evaporation, which is unique in that the deposition is effected under nonequilibrium conditions (Figure 5.25).

Figure 5.25 Illustration of the concept of vapor injection source technology (VIST), an application of flash evaporation.

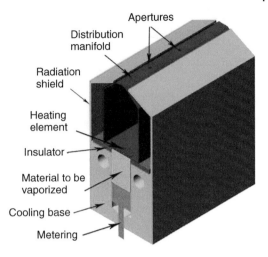

The VIST method is applied to mitigate the trade-off between high deposition rate and material consumption, which is unavoidable with the conventional evaporation technologies. Material evaporates instantaneously on physical contact with the heated region.

In normal OLED manufacturing using conventional evaporation sources, OLED material is heated in one evaporation source for a long period, for example, 1 day or 1 week. Therefore, material decomposition is often a serious problem.

In the VIST method, powder materials transferred to the heated region are exposed to high temperatures for a much shorter time period (e.g., ~2 s), so the degree of thermal decomposition is negligible. Also, the material is maintained at room temperature before it is evaporated, so the stored material does not decompose whatsoever.

Figure 5.26 compares the performance levels of white-emission OLED devices using point-source technology (0.2 nm/s) versus VIST technology (5 nm/s) using thermally sensitive blue emission material. OLED device efficiency and voltage are almost equal, and the OLED device lifetime using VIST technology is longer than that using point source. Thus, VIST is more effective for thermally sensitive materials. In particular, many thermally sensitive materials are available on the market, so the demand for VIST may increase.

Many other high-material-utilization sources have been proposed, but doping capability is problematic for many of them. This is because most of the evaporation using these sources is controlled by equilibrium conditions, whose doping condition is subjected to the partial pressure of the organic materials to be evaporated. With VIST, most of the evaporation takes place under

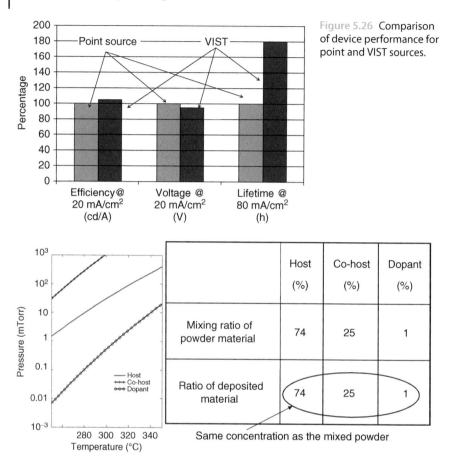

Figure 5.26 Comparison of device performance for point and VIST sources.

Figure 5.27 Comparison of doping performance for materials with widely different evaporation pressures.

nonequilibrium conditions. This problem can be avoided by mixing the powder materials in advance, to control the doping concentration, even for multiple materials with totally different evaporation pressures (Figure 5.27).

The VIST method (Figure 5.25) achieves uniform deposition using a gas transfer structure called a *manifold*, which is equivalent of the showerhead in CVD equipment. The manifold has multiple apertures, which connects the manifold to the material supply component, called an *injector*. As shown in Figure 5.28, it is possible to connect multiple injectors to one manifold and perform multilayer deposition using only one manifold.

There are several ways to implement the VIST method in actual equipment, such as installing multiple linear manifolds (Figure 5.29a) to produce inline

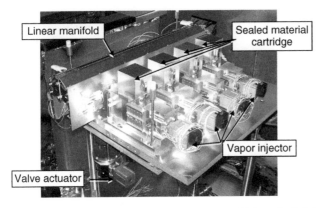

Figure 5.28 A dopant-monitoring apparatus in which multiple injectors are connected to a single manifold.

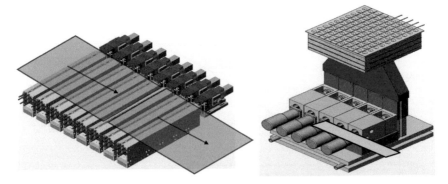

Figure 5.29 Examples of multilayer fabrication using a VIST source.

equipment or introducing an area (evaporation) manifold to deposit multiple layers in one chamber with multiple injectors connected to one manifold (Figure 5.29b). In either case, high material utilization would be possible.

One advantage of the VIST method is that the equipment can be operated without a crystal thickness monitor (discussed in Section 3.2.2.3), which requires frequent high-cost maintenance. The VIST method employs a Pirani gauge attached to the manifold to monitor pressure (Figure 5.30). Conventional vacuum evaporation used for the OLED manufacturing actually measures the deposition rate from the deposited film thickness change by monitoring the oscillation frequency of the quartz substrate (see Section 3.2.2.3 for detail). Pirani gauge directly can measure the pressure to know the deposition rate. As shown in Figure 5.31, the crystal thickness monitor and the Pirani gauge are closely correlated. As the deposition rate measurement by the Pirani gauge does not require parts exchange during manufacturing that are different

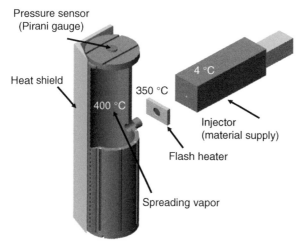

Figure 5.30 A pressure-monitoring apparatus in which a Pirani gauge is attached to a manifold.

Pressure sensor (Pirani gauge)

Heat shield

350 °C

400 °C

4 °C

Injector (material supply)

Flash heater

Spreading vapor

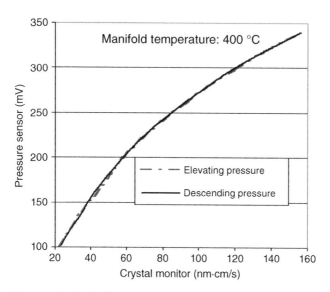

Figure 5.31 Graph illustrating correlation between crystal thickness monitoring and pressure sensing.

from those involved with the crystal monitor, the equipment utilization ratio can be increased. As shown in Figure 5.27, the doping ratio using the VIST method is determined by the mixture ratio of dopant in the powder, so the dopant concentration can be calculated from the total thickness, which can be calculated from the total pressure measured by the Pirani gauge.

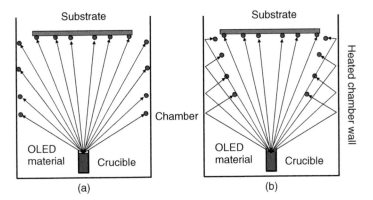

Figure 5.32 Schematic representations of OLED material deposition mechanism using (a) conventional evaporation and (b) hot-wall methods.

5.3.2 Hot-Wall Method

As discussed in Section 3.2.2.1, a nonnegligible amount of OLED material is actually deposited on the substrate, and most of the volume is deposited elsewhere, such as on a chamber wall, shutter, or control plate, as shown in Figure 5.32a. Because of the low utilization of OLED material, the material cost during manufacturing is both expensive and problematic.

The hot-wall method is used to avoid material deposition upon evaporation by heating the chamber wall during evaporation to achieve high material utilization (Figure 5.32b).

Figure 5.33 shows the variation in behavior of an incident molecule that hits a wall when the wall temperature changes.

- If the temperature is low (Figure 5.33a), the incident molecule hits the wall and then adheres to it.
- If the temperature is higher, the incident molecule has more kinetic energy, so after it hits the wall, it detaches from it (Figure 5.33b).
- If the temperature is much higher, the incident molecule does not adhere to the wall but makes an elastic collision (Figure 5.33c).

Figure 5.33 Schematic depiction of the behavior of an incident evaporation molecule on collision with a wall.

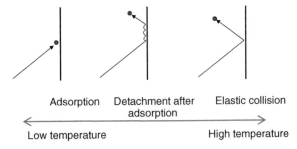

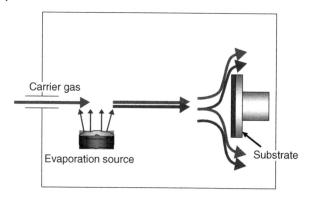

Figure 5.34 Flowchart showing thin-film deposition using the organic vapor-phase deposition (OVPD®; a trademark of Universal Display Corp.) method.

The hot-wall method has high material utilization but has drawbacks such as the following:

- Damage to OLED device due to radiant heat
- Deformation of shadow mask due to radiant heat
- Necessity for enough outgassing after maintenance to avoid the impurities coming out from the chamber wall (grease, O-ring, etc.) during the deposition process.

5.3.3 Organic Vapor-Phase Deposition (OVPD) Method

The OVPD method consists of transferring material via a carrier gas and depositing film on the cooled substrate from the shower plate (Figure 5.34) [12].

Figure 5.35 shows a simplified example of actual OVPD equipment. Material consumption can be switched on and off by valve operation, thus ensuring good material utilization as unnecessary evaporation can be suppressed, such as during substrate-handling, startup, and cooldown periods.

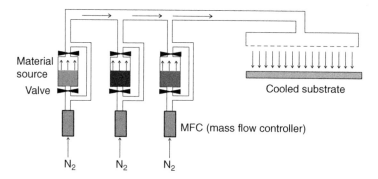

Figure 5.35 Schematic representation of the OVPD apparatus.

Short Thermal Exposure source (STEx) further improves the merit of OVPD equipment. It maintains the organic material in a room temperature reservoir, and carrier gas with controlled concentration of organic material is made by means of flash vaporization of material supplied by aerosol method [13].

References

1 M. Kim, M. Song, S. Lee, H. Kim, J. Oh, and H. Chung, Control of emission zone in a full color AMOLED with a blue common layer, *SID 2006 Digest*, pp. 135–138 (2006).

2 T. Matsumoto, T. Yoshinaga, T. Higo, T. Imai, T. Hirano, and T. Sasaoka, High-performance solution-processed OLED enhanced by evaporated common layer, *SID 2011 Digest*, pp. 924–927 (2011).

3 T. Tsujimura, T. Hakii, T. Nakayama, H. Ishidai, T. Kinoshita, S. Furukawa, K. Yoshida, and K. Osawa, Development of a color-tunable polychromatic organic-light-emitting-diode device for roll-to-roll manufacturing, *J. Soc. Inf. Display* **24**(4):262–269 (2016).

4 J. Kido, M. Kimura, and K. Nagai, Multilayer white light-emitting organic electroluminescent device, *Science* **267**:1332–1334 (1995).

5 C. Hosokawa, M. Eida, M. Matsuura, K. Fukuoka, H. Nakamura, and T. Kusumoto, Organic multi-color electroluminescence display with fine pixels, *Synth. Metals* **91**(1–3):3–7 (1997).

6 S. Lee, B. Chin, M. Kim, T. Kang, M. Song, J. Lee, H. Kim, H. Chung, M. Olk, E. Bellmann, J. Baetzold, S. Lamansky, V. Savvateev, T. Hoffend, J. Staral, R. Roberts, and Y. Li, Novel patterning method for full-color organic light-emitting devices: laser induced thermal imaging (LITI), *SID 2004 Digest*, pp. 1009–1011 (2004).

7 M. Boroson, L. Tutt, K. Nguyen, D. Preuss, M. Culver, and G. Phelan, Non-contact OLED color patterning by radiation-induced sublimation transfer (RIST), *SID 2005 Digest*, pp. 972–975 (2005).

8 M. Burghart, A. Dutkoowiak, L. Tandler, J. Richter, G. Haasemann, H. Gross, U. Seyfert, T. Strietzel, and C. Hecht, High resolution vacuum patterning of organic- and metal-layers for organic electronic devices, *SID Symposium Digest of Technical Papers*, Vol. **44**, Issue 1, pp. 427–430, (2013).

9 H. Becker, S. Heun, K. Treacher, A. Büsing, and A. Falcou, Materials and inks for full-colour PLED displays, *SID 2002 Proceedings*, pp. 780–782 (2002).

10 M. O'Regan, D. Lecloux, C. Hsu, E. Smith, A. Goenage, and C. Lang, Solution processing of small molecule OLED materials at DuPont, *IMID/IDMC 2006 Digest*, p. 1689 (2006).

11 J. P. Markham, S.-C. Lo, T. D. Anthopoulos, N. H. Male, E. Balasubramaniam, O. V. Salata, P. L. Burn, and I. D. W. Samuel, Highly efficient solution-processible phosphorescent dendrimers for organic light-emitting diodes, *J. SID* **11**(1):161 (2003).

12 M. Schwambera, N. Meyer, S. Leder, M. Reinhold, M. Dauelsberg, G. Strauch, M. Heuken, H. Juergensen, T. Zhou, T. Ngo, J. Brown, M. Shtein, and S. R. Forrest, Modeling and fabrication of organic vapor phase deposition (OVPD) equipment for OLED display manufacturing, *SID 2002 Digest*, p. 894 (2002).

13 M. Gersdorff, M. Long, D. Keiper, M. Kunat, B. Gopi, C. Cremer, B. Beccard, and M. Schwambera, Enabling high throughput OLED manufacturing by carrier gas enhanced organic vapor deposition (OVPD), *SID Symposium Digest of Technical Papers*, Vol. **42**, Issue 1, pp. 516–519 (2011).

6

TFT and Driving for Active-Matrix Display

Thin-film transistor technologies for active-matrix driving are discussed in this chapter.

6.1 TFT STRUCTURE

Figure 6.1 shows structural differences between polysilicon (polycrystalline silicon) and amorphous silicon (sometimes abbreviated as a-Si or α-Si).

As shown in the figure, amorphous silicon has an irregular structure, so the potential energy is irregularly distributed. Therefore, the moving charge is scattered and mobility is low. The binding state fluctuates as a result of trapping and detrapping, so TFT characteristics change over time. On the other hand, polysilicon inside a grain has a cubic crystalline structure similar to that of single-crystal silicon, where the charges move within a periodical potential in the same way as in a single crystal, so the mobility is very high and stable.

Figure 6.2 shows a TFT substrate used in an actual active-matrix OLED display product. It is very similar to the LCD display TFT backplane except for the existence of a driver TFT in the OLED backplane.

There are two basic types of TFT structure, according to the electrode configuration: top-gate and bottom-gate TFT [1]. When the gate electrode is located above the source–drain electrode, it is called *top-gate* (or *staggered*) *TFT*. When the source and drain are above the gate electrode, it is called *bottom-gate* (or *inverted staggered*) *TFT*. Figure 6.3 shows examples of top- and bottom-gate TFT structures. Advantages of top-gate structure include the following:

- Low-resistance wiring can be easily used because gate wiring, which is subjected to high temperatures during fabrication, is formed after high-temperature processes such as polysilicon formation or gate insulator deposition.

OLED Display Fundamentals and Applications, Second Edition. Takatoshi Tsujimura.
© 2017 John Wiley & Sons, Inc. Published 2017 by John Wiley & Sons, Inc.

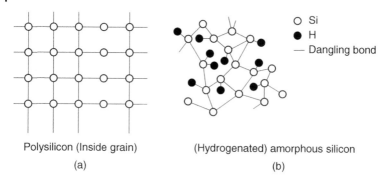

Polysilicon (Inside grain)
(a)

(Hydrogenated) amorphous silicon
(b)

○ Si
● H
— Dangling bond

Figure 6.1 Differences between (a) polysilicon (inside grain) and (b) (hydrogenated) amorphous silicon.

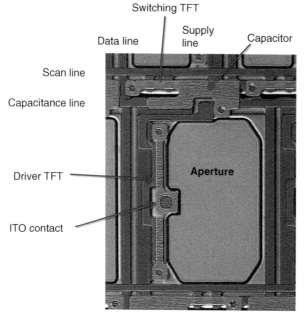

Switching TFT

Data line

Supply line

Capacitor

Scan line

Capacitance line

Driver TFT

Aperture

ITO contact

Figure 6.2 An example of a polysilicon-driven OLED display panel.

- A self-aligned doping process is available using gate metal as a mask. Self-aligned structure has a merit in reduced parasitic capacitance, which causes charging errors of the driver TFT gate node. Also accurate positioning of the doping region by self-alignment has merits in the N-region positioning control so that the hot carrier effect caused by the gate to the source electric field can be reduced.

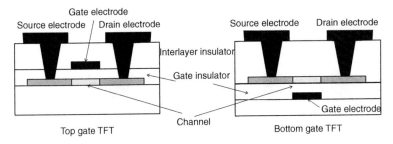

Figure 6.3 (a) Top-gate and (b) bottom-gate TFTs.

Advantages of bottom-gate TFT structure include the following:

- A low-cost amorphous silicon TFT (Figure 6.5) facility can be used for fabrication.
- Continuous deposition can be applied in a vacuum environment between the gate insulator and silicon, which provides better and more stable TFT performance.

Here, we discuss the OLED backplane, using a top-gate TFT structure.

Figure 6.4 shows a cross section of a polysilicon TFT, with an N-channel doping region and a P-channel doping region on the same substrate. This is called a complementary metal oxide semiconductor (CMOS) circuit, the main advantage of which is low-power driving (discussed in Sections 6.4.4 and 6.4.5). However, as shown in Table 6.1, CMOS process flow is more complicated than that of amorphous silicon (Figure 6.5), which has only four or five patterning steps. Thus, there is a trade-off between cost reduction by circuit integration and cost increase due to increased process complexity.

To address this issue, many companies are developing CMOS mask step reduction and single-channel (either PMOS or NMOS) circuit integration technologies (see the discussion in Section 6.4.5).

6.2 TFT PROCESS

The TFT fabrication process for OLED application is discussed in this section.

6.2.1 Low-Temperature Polysilicon Process Overview

Figure 6.6 presents one example of the polysilicon TFT fabrication process.

After amorphous silicon deposition by PECVD (plasma-enhanced chemical vapor deposition) and its dehydrogenation, the Si film is crystallized by

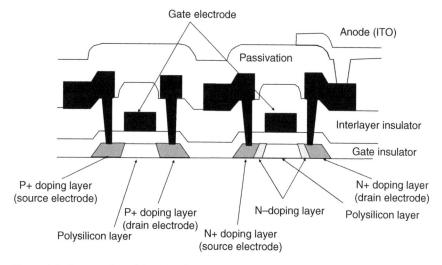

Gate electrode

Anode (ITO)

Passivation

Interlayer insulator

Gate insulator

P+ doping layer
(source electrode)

N+ doping layer
(drain electrode)

P+ doping layer
(drain electrode)

N–doping layer

Polysilicon layer

Polysilicon layer

N+ doping layer
(source electrode)

Figure 6.4 Cross section of CMOS polysilicon TFT.

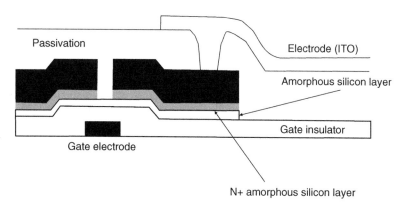

Passivation

Electrode (ITO)

Amorphous silicon layer

Gate insulator

Gate electrode

N+ amorphous silicon layer

Figure 6.5 Cross section of typical amorphous silicon TFT.

excimer laser annealing (ELA) (Figure 6.6a). After polysilicon patterning, the gate insulator and gate electrode are formed (Figure 6.6b). The source–drain contact is formed by an ion doping process using the gate electrode as a mask. For the N channel, leakage current is a serious issue, as discussed in Section 6.4.1, so a lightly doped drain (LDD) region is formed by an additional low-concentration doping process using an offset created by anodic oxidation of the gate electrode or the gate electrode etching bias (Figure 6.6c). After an interlayer dielectric film is formed, contact holes are made, the source–drain layer is formed (Figure 6.6d), and then an ITO pattern, used for an OLED anode electrode, is fabricated to produce a TFT substrate.

Table 6.1 Typical Steps in the CMOS Polysilicon TFT Process

Photoengraving Process (PEP) Step	Activity or Procedure
.	Cleaning CVD (a-Si) Dehydrogenation annealing Ion doping (Br channel doping)
1	Si island resist patterning Dry etching
2	P doping mask formation Ion doping (P-channel doping) CVD (gate insulator) Sputter (gate electrode)
3	Gate electrode mask formation Dry etching Ion doping (LDD doping by P)
4	Br doping mask (P^+) formation Ion doping (Br)
5	P doping mask (N^+) formation Ion doping (P) Activation anneal
6	Contact hole mask formation Wet etching Sputter (data line)
7	Data line mask formation Wet etching CVD (passivation)
8	Passivation patterning mask Dry etching H_2 forming anneal Sputter (ITO)
9	Pixel mask formation Wet etching Inspection

Table 6.1 lists steps in a process flow for the fabrication of a CMOS backplane (a CMOS TFT circuit contains both P- and N-channel devices). In this embodiment, nine mask patterning and five doping steps are required; however, as complicated process steps result in cost increase, simplified processes for patterning and doping are employed for state-of-the-art fabrication processes.

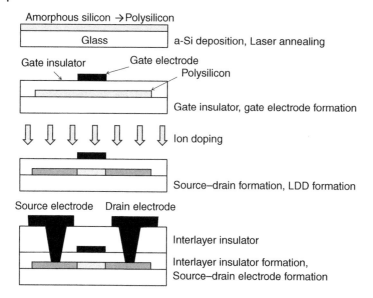

Figure 6.6 Flowchart of polysilicon TFT process.

6.2.2 Thin-Film Formation

Films used for TFT are generally prepared by two kinds of deposition methods: CVD and sputtering (also called PVD: physical vapor deposition). The films prepared by CVD for TFTs are amorphous silicon for the channel layer (polysilicon is formed by annealing an amorphous silicon film), N$^+$ amorphous silicon for the contact layer (although conventional polysilicon TFT has been using the doping implantation to form the contact region just like semiconductor process, the use of N$^+$ amorphous silicon, popularly used for amorphous silicon TFT process, is increasing to reduce mask steps), gate insulator, an interlayer dielectric, and a passivation film such as silicon oxide (SiO$_x$), silicon nitride (SiN$_x$), and silicon oxynitride. In particular, PECVD, a method involving the deposition of material created by reactions in a plasma of a gas mixture, for example, silane (SiH$_4$) and hydrogen (H$_2$) in the case of amorphous silicon deposition, are often employed (Figure 6.7a).

On the other hand, metal films such as aluminum, aluminum alloys (AlNd is often used for gate bus wiring), molybdenum, molybdenum alloys (such as MoW and MoTa), chromium, tantalum, and copper are deposited by the sputtering method. Insulators such as SiO$_2$ or a transparent electrode such as ITO or IZO (indium zinc oxide) can also be deposited by radiofrequency (RF) sputtering or the ion plating method. The sputtering (Figure 6.7b) method deposits onto a substrate, material ejected from the sputter target by recoil under impact by energetic bombardment of an ionized gas such as argon. Films deposited by

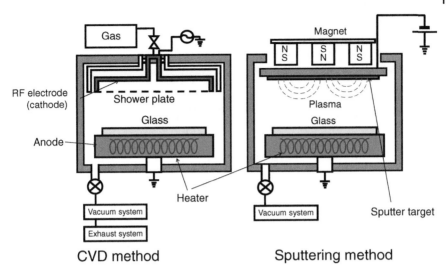

Figure 6.7 Schematic representation of equipments used to form TFT by thin-film deposition: (a) chemical vapor deposition (CVD) method; (b) sputtering method.

the sputtering method have good adhesion properties because the material is deposited with high kinetic energy. As discussed in Section 2.3.7.4, ITO used for anode electrodes needs workfunction control to have smooth hole injection into the OLED device. The workfunction of ITO needs to be close to the HOMO energy level of the organic material, which receives holes. ITO surface modification, such as oxygen plasma, UV-O_3 treatment, or CF_x plasma treatment, is made to have a workfunction deep enough to inject holes. Also, a liquid solution treatment is used to clean the surface and to deepen the workfunction to match with the HOMO of the organic layer. The roughness of the ITO is also an important factor. As the films of organic materials used to make OLEDs are very thin, the ITO needs to have a very flat and smooth surface. The roughness of ITO used for OLED is usually less than a few nanometers measured by AFM. To achieve this, some manufacturers use polished ITO. Amorphous ITO deposition and successive annealing is also used to achieve a smooth surface.

6.2.3 Patterning Technique

In Section 6.2.1, the film formation process was discussed. The deposited film is patterned by a set of processes called PEP (photoengraving process) as shown in Figure 6.8.

Photoresist resin is coated on the deposited film by spin coating or slit coating. The photoresist type that is removed by photoexposure is called *positive-type resist*. On the other hand, the photoresist that remains following photoexposure is called *negative-type resist*. Positive-type resist is more

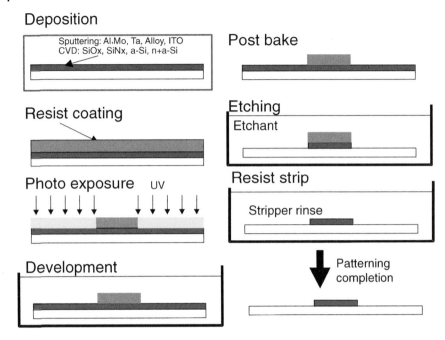

Figure 6.8 Flowchart of patterning process (photoengraving process [PEP]) for TFT fabrication.

popular than negative-type resist because a particle on the negative-type resist causes removal of the film underneath the particle in the etching process and results in an electrical short between the layers, while positive-type does not, and also because the pattern dimension does not change very much and remains accurate in terms of positive type. (Negative-type resist is used in color filters and organic black matrix layers, as such high-optical-density films cannot be sufficiently photoexposed to be removed if the resist is of the positive type.)

Most positive-type photoresist uses naphthoquinone diazide (NQD) as the photoreactive agent. Novolac resin and NQD, which are contained in positive-type photoresist, do not dissolve in an alkaline solution (the presence of NQD inhibits dissolution of the Novolac polymer due to hydrogen bonding between Novolac and NQD); however, with UV irradiation, NQD is transformed to indene carboxylic acid, which dissolves in an alkaline solution (containing, e.g., tetramethylammonium hydroxide [TMAH]) and the photoresist is patterned (Figure 6.9).

Figure 6.10 shows photoexposure equipment for resist patterning. A mercury lamp is used as a light source for the exposure. There are two types of exposure equipment: one that exposes the whole substrate simultaneously and

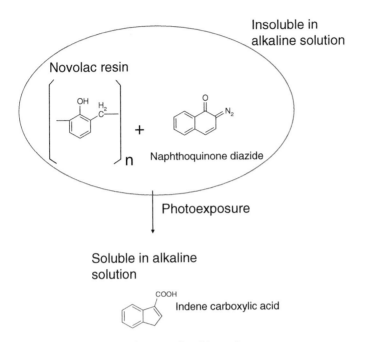

Figure 6.9 Reaction mechanism of positive resist.

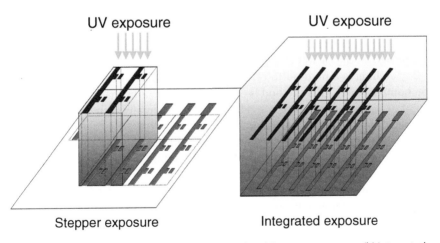

Figure 6.10 Photoexposure tools for resist patterning: (a) stepper exposure; (b) integrated exposure.

Resist coating

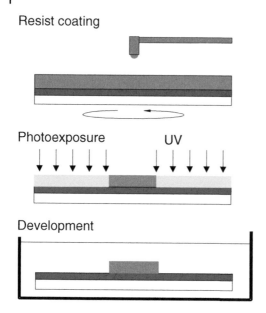

Figure 6.11 Flowchart of resist pattern formation process: (a) resist coating; (b) photoexposure; (c) development.

Photoexposure   UV

Development

another that exposes areas consecutively. A mercury-vapor lamp emits line spectra including G-line, H-line, and I-line. The G-line (436 nm) is the most popular, but other wavelengths are also used. For the planarization layer and color filter patterning, a broad spectrum containing the G-, H- (405 nm), and I-lines (365 nm) is used. For fine patterning or planarization-layer patterning, the I-line is useful.

After the photoexposure (Figure 6.11a and b) stages, the target resist pattern is immersed in an alkaline solution. The film area with no photoresist coating is removed by dry or wet etching, and the film pattern is formed (Figure 6.12).

Dry etching is a method to remove a film region by chemical reaction and ion bombardment. The method is carried out in vacuum chamber and is capable of defining small feature size. However, it is normally higher cost than wet etching due to high tool cost and low throughput. It also has demerits such as poor selectivity (discussed later) and potential radiation damage. In the case of wet etching, chemical reaction etches off the film and is carried out in the bath placed in the atmospheric environment. It is lower cost than dry etching. Higher etching rate and good selectivity for most materials can be obtained. As it is isotropic etching, it is not adequate for defining small feature size [2].

Commonly used gases for dry etching are as follows [2]:

- CF_4, SF_6, and $BCl_2 + Cl_2$ for silicon
- $CHF_3 + O_2$ and $CF_4 + H_2$ for SiO_2
- $-CF_4 + O_2$ (H_2) or CHF_3 for Si_3N_4.

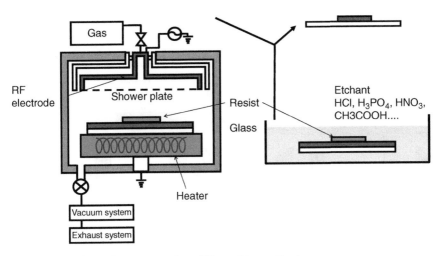

Figure 6.12 Schematic representation of film etching methods.

Commonly used etchants for wet etching are as follows [2]:

- HF and BHF for SiO_2
- HF, BHF, and H_3PO_4 for Si_3N_4
- $HCL + H_2O$ and NaOH for Al.

It is important to select the appropriate dry-etching gas (e.g., CF_4 gas etches Si_3N_4 but does not etch aluminum) or appropriate wet etchant so that the etching rate of the film to be etched is large enough compared with the etching rate of the layer beneath in order to pattern the film without damage to the underlying layer. (The etching speed differential between the layers is called the *etch selectivity*.) Photoresist is also used as a planarization layer as shown in Figure 6.13. After TFT formation, the surface becomes bumpy due to the film pattern. However, OLED films to be deposited on it are very thin (a few tens of nanometers), so the bumpy surface profile can cause OLED film pinholes, which causes an electric short between cathode and anode. A planarization layer is used to help circumvent this issue. A planarization layer also reduces the parasitic capacitance between electrode and TFT wiring so that the aperture ratio can be increased by shortening the electrode to wiring distance.

6.2.4 Excimer Laser Crystallization

Several different crystallization methods for silicon film are used, depending on how energy is applied to the film. One approach consisting of crystallization of silicon film by annealing is called *solid-phase crystallization* (SPC). As shown in Figure 6.14, the crystal grain grows from the crystalline seed, which serves as a growth center.

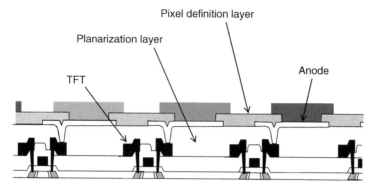

Figure 6.13 An example of a planarization layer for active-matrix OLED display pixels.

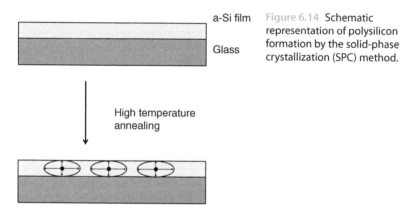

Figure 6.14 Schematic representation of polysilicon formation by the solid-phase crystallization (SPC) method.

As the use of glass substrates for OLEDs or LCDs is problematic at high temperatures, for example, strain point of 669 °C for Eagle-XG® glass substrate, while solid-phase crystallization temperature of noncrystalline silicon is over 700 °C [3]), several technologies have been developed to decrease the crystallization temperature. One of the most popular approaches is laser annealing [4, 5], which provides energy by absorption of a laser beam within the silicon film to decrease the effective temperature of the glass substrate when Si is molten. Another popular approach is metal-induced crystallization (MIC), which decreases the crystallization temperature by using nickel as a crystallization seed that has the same body-centered cubic (BCC) lattice as does a silicon crystal (see Section 6.1). These are termed *low-temperature polysilicon* (LTPS) processes, as the temperature is lower than that of the normal crystallization of silicon.

ELA is the most popular among all the low-temperature crystallization processes. An excimer laser irradiates the silicon film by transmitting a pulsed (several tens of nanoseconds) light beam of 308 nm wavelength generated by

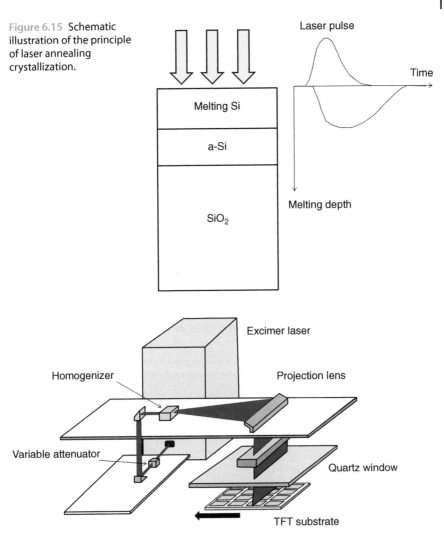

Figure 6.15 Schematic illustration of the principle of laser annealing crystallization.

Figure 6.16 Schematic representation of apparatus used for excimer laser annealing crystallization.

XeCl, which increases the local silicon film temperature to above the melting point of silicon (Figures 6.15 and 6.16). As the film is locally heated, glass substrates that cannot withstand temperatures above 600 ~ 700 °C can be used.

Figure 6.17 shows the dependence of laser irradiation energy on crystal grain size due to ELA. As the irradiation energy increases, grain size also increases (from region A to B); however, at a certain energy level, the crystal grains acquire a fine granular shape (in region C). This is because the mechanism

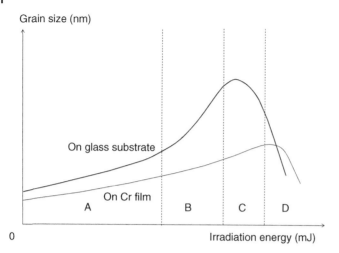

Figure 6.17 Graphical depiction of dependence of laser irradiation energy on crystal grain size.

changes in response to the supercooling process, in which the silicon film maintains itself in the molten state even after the temperature drops below the solidification point. In this state, as the temperature decreases, solidification suddenly begins. TFT characteristics such as mobility and S value (subthreshold slope) are improved as the grain size increases in region B, as shown in Figure 6.18. To ensure adequate performance, it is necessary to maintain region B conditions during manufacturing.

In terms of the laser, energy density is not uniform in a beam. In particular, the energy at the beam edge tends to differ widely (lower energy) from that at the center. To avoid any change in TFT characteristics at the beam edge, the beam location is shifted for every pulse to form an overlap region (e.g., 90% overlap) with the previous pulse. As a result, the same location receives multiple pulses in short time. A multibeam shot provides the advantage of grain size enlargement (Figure 6.19) but with the trade-off of lower manufacturing throughput. (To avoid this problem, the sequential lateral solidification [SLS] method has been introduced. The method is discussed in Section 6.5.1.1.)

6.3 MOSFET BASICS

Figure 6.20 shows transistor behavior in linear region operation [6]. A transistor that switches on and off depending on the gate electrode potential is called a field effect transistor (FET). Especially, an FET whose channel conductance is modulated by the gate electrode placed across the dielectric layer is called a

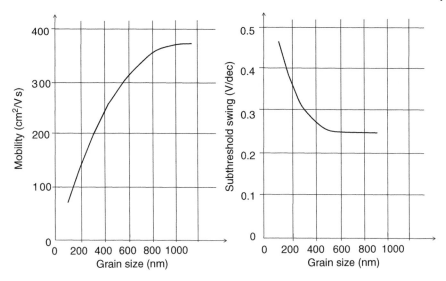

Figure 6.18 Plots showing correlation between crystal grain size and TFT characteristics.

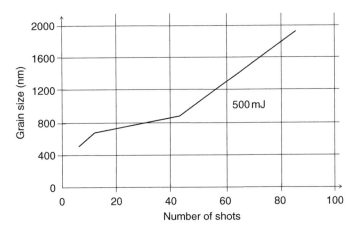

Figure 6.19 Graphical representation of correlation between laser shot number and crystal grain size.

metal oxide semiconductor field effect transistor (MOSFET) or metal insulator semiconductor field effect transistor (MISFET).

When V_{GS} (gate–source voltage; V_G–V_S) is higher than the threshold voltage (V_{th}), electric current is generated by V_{DS} (drain–source voltage; V_D–V_S). Here, we assume that $V_S = 0$ (source potential) and $V_B = 0$ (substrate potential). Also, the potential at coordinate y relative to the source potential is assumed to be $V_C(y)$. It is also assumed that V_{th} is the same in all locations in the channel

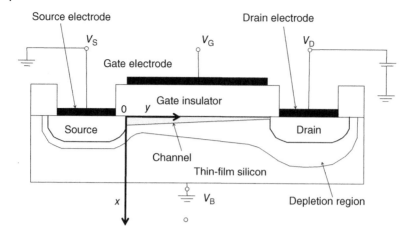

Figure 6.20 A schematic diagram of N-channel transistor operation in the linear region.

and that the y-axis electric field is dominant and the x-axis electric field is minor. This assumption is called *gradual-channel approximation.*

The following condition is also assumed:

$$V_{GS} > V_{th} \tag{6.1}$$

$$V_{GD} = V_{GS} - V_{DS} \geq V_{th} \tag{6.2}$$

The boundary condition can be expressed as

$$V_C(y = 0) = V_S = 0 \tag{6.3}$$

$$V_C(y = L) = V_{DS} \tag{6.4}$$

When the total charge in the channel is $Q(y)$, then

$$Q(y) = -C_{Ox}\{V_{GS} - V_C(y) - V_{th}\} \tag{6.5}$$

If surface mobility is μ, the resistance dR caused by a small region dy in the channel can be expressed as follows:

$$dR = -\frac{dy}{W \mu Q(y)} \tag{6.6}$$

$$dV_C = I_D dR = -\frac{I_D \, dy}{W \mu Q(y)} \tag{6.7}$$

Integrating from $y = 0$ to $y = L$ (channel length), we obtain

$$\int_0^L I_D \, dy = -W \mu \int_0^{V_{DS}} Q(y) dV_C \tag{6.8}$$

So

$$I_D = \frac{\mu C_{Ox}}{2} \frac{W}{L} \{2(V_{GS} - V_{th})V_{DS} - V_{DS}^2\} \tag{6.9}$$

Thus, the drain current in the linear region ($V_{DS} < V_{GS} - V_{TH}$) can be expressed as

$$I_D = \mu C_{OX} \frac{W}{L} \left\{ (V_{GS} - V_{th})V_{DS} - \frac{V_{DS}^2}{2} \right\} \tag{6.10}$$

Here, μ is mobility, V_{th} is threshold voltage, C_{OX} is gate insulator capacitance per unit area, and W and L are channel width and channel length, respectively.

On the other hand, when V_{DS} is larger than $V_{GS} - V_{th}$, the channel disappears at the pinchoff point and the drain current does not increase, even with larger V_{DS} (saturation region operation). The current is equal to the current of

$$V_{DS} = V_{GS} - V_{th} \tag{6.11}$$

so, substituting Eq. (6.11) for Eq. (6.9), we can express the drain current in the saturation region as

$$\begin{aligned} I_D &= \frac{\mu C_{Ox}}{2} \frac{W}{L} \{2(V_{GS} - V_{th})(V_{GS} - V_{th}) - (V_{GS} - V_{th})^2\} \\ &= \frac{\mu C_{Ox}}{2} \frac{W}{L} (V_{GS} - V_{th})^2 \end{aligned} \tag{6.12}$$

6.4 LTPS-TFT-DRIVEN OLED DISPLAY DESIGN

In this section, we discuss methods for designing a high-quality display with OLED devices.

6.4.1 OFF Current

One problem with an LTPS-TFT is high leakage current. The leakage current is caused by insufficient charge blocking capability of the contact region formed between the intrinsic region (nondoping region; also called the *i*-layer) and the doped region created by ion doping.

As shown in Figure 4.33, most pixel circuits contain a switching TFT, which relays the grayscale signal to the pixel, and a driver TFT, which causes electric current to flow to the OLED device for emission. The driver TFT does not need very low leakage current; however, if the switching TFT has a high leakage current, the pixel capacitor cannot retain the signal level for accurate OLED device emission, and an abnormal display image or crosstalk may result.

To suppress the leakage current of TFTs, an alternative method is to employ a double-gate structure [7] (Figure 6.21) that connects two TFTs in series or a multigate structure that connects multiple TFTs in series. (*Note*: The term *double-gate structure* can also refer to a TFT containing both top-gate electrode and bottom-gate electrodes; this term is confusing, so we must note the context when we find it.)

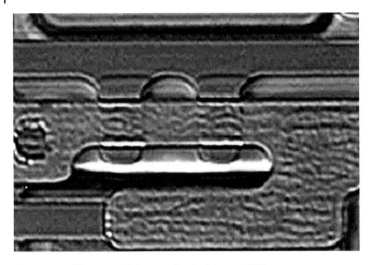

Figure 6.21 Photographic image of a double-gate TFT structure.

Also, as an N-channel LTPS-TFT normally has a high leakage current, a lightly doped drain (LDD) structure such as that shown in Figure 6.4 is often used [8]. Figure 6.22 plots the correlation between LDD length and TFT characteristics. As the LDD length increases, the OFF current is largely decreased. However, the ON current also decreases proportionally with the increase in the LDD length, so this is potentially a trade-off situation. Optimum LDD length should be chosen for the best design. LDD structure is also effective to avoid the reliability problem caused by the hot carriers, induced by the strong electric field between gate and drain. By applying an N-doping layer, the gate to drain electric field is weakened so that the damage due to hot carriers is diminished.

6.4.2 Driver TFT Size Restriction

An OLED device is driven by electric current, and to satisfy the luminance specification with respect to visibility in ambient environment conditions, a large current is necessary. Here, we explain the TFT characteristics necessary for this purpose.

The electric current necessary to drive a pixel can be expressed as

$$I_{\text{pixel}} = \frac{L_{\max} \times 9a^2}{\eta} \tag{6.13}$$

where $L_{\max}$ is the maximum display luminance, I_{pixel} the maximum current that flowing through OLED diode, η the current efficiency of OLED diode, and a the subpixel pitch.

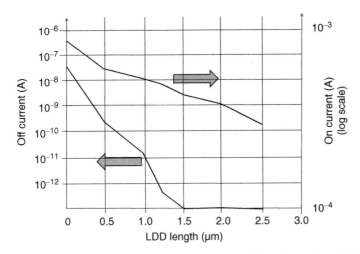

Figure 6.22 Graph depicting correlation between lightly doped drain (LDD) length and TFT characteristics.

According to Eq. (6.12), maximum current at a certain drain voltage can be described as

$$I_{pixel} = \frac{W}{2L}\mu C_{Ox}(V_{GS} - V_{th})^2 \tag{6.14}$$

where W is the driver TFT channel width, L the driver TFT channel length, μ the TFT mobility, C_{OX} the TFT channel capacitance, V_{GS} the TFT gate voltage, and V_{th} the threshold voltage.

Therefore,

$$W = \frac{18L_{max}a^2L}{\eta\mu C_{Ox}(V_{GS} - V_{th})^2} \tag{6.15}$$

As W increases, the aperture ratio decreases in the bottom emission device, so OLED degrades more rapidly because of an increase in the current density flowing through the OLED device. An excessively high W value is considered to be disadvantageous for an OLED display, as large TFT creates much overlap region between electrodes, which increases the probability of failure due to short defects. To ensure good yield, it is necessary to design a TFT whose size is not unwieldy compared with amorphous silicon TFT-LCD, which uses a narrow channel width (e.g., $W = 20\,\mu m$).

6.4.3 Restriction Due to Voltage Drop

As an OLED is a current-driven device, it is very important to take voltage drop (Figure 6.23b) into account in designing a large display.

(a)

(b)

Figure 6.23 Images of displays (a) without voltage drop and (b) with severe voltage drop.

Current flowing in a supply line can be described as

$$I_{\text{supply}} = 3mI_{\text{pixel}} \tag{6.16}$$

where m is the number of pixels along the horizontal direction.

Wiring resistance can be expressed as

$$R_{\text{supply}} = \frac{3ma\rho_{\text{supply}}}{d_{\text{supply}}W_{\text{supply}}} \tag{6.17}$$

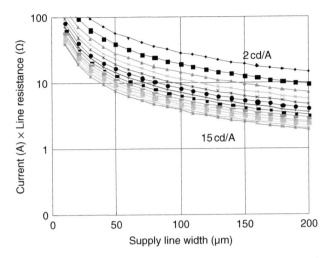

Figure 6.24 Voltage drop simulation results, assuming wiring with same parameters as those in LCDs.

where ρ_{supply} is the resistivity of supply line, d_{supply} the thickness of supply line, and W_{supply} the width of supply line.

Then the maximum voltage drop that can occur during display driving is

$$V_{drop} = I_{supply}R_{supply} = 9m^2 I_{pixel}a\frac{\rho_{supply}}{d_{supply}W_{supply}} \tag{6.18}$$

If standard resistivity value, thickness, and line width used for the LCD display are assumed, then a serious voltage drop such as that graphed in Figure 6.24 can occur. For example, if $\rho_{supply} = 4\,\mu\Omega\cdot cm$, $W_{supply} = 5\,\mu m$, $d_{supply} = 200\,nm$, and $I_{pixel} = 10\,\mu A$, then the maximum voltage drop along the supply line can be calculated as $V_{drop} = 4.22\,V$ (as shown in Figure 6.23), which is a serious problem for displaying images.

To avoid problems related to voltage drop, OLED efficiency improvement, such as the use of a phosphorescent emitter, an outcoupling enhancement, or low-resistance wiring, consisting of aluminum or copper at low impurity concentrations, is necessary.

The use of low-impurity wiring material, increase in wiring thickness with improved gate insulator coverage, increase in wiring width, and an improved driving technique can reduce the voltage drop and provide sufficient luminance uniformity, as shown in Figures 6.25 and 6.26. In particular, to achieve low-resistance wiring, taper angle control is very important in order to ensure gate insulator coverage for high production yields.

Using Eq. (6.18), we obtain

$$V_{drop} = I_{supply}R_{supply} = 9m^2 I_{pixel}a\frac{\rho_{supply}}{d_{supply}W_{supply}} \tag{6.19}$$

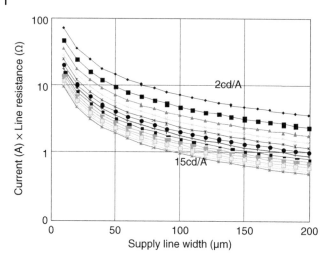

Figure 6.25 Plot showing voltage drop reduction with increase in wire thickness.

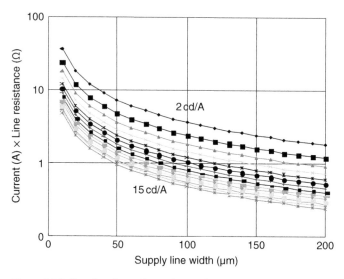

Figure 6.26 Plot showing voltage drop reduction with increase in wire thickness in a low-resistivity wiring material.

Using Eq. (6.13), we obtain

$$I_{pixel} = \frac{L_{max} \times 9a^2}{\eta} \tag{6.20}$$

Therefore, voltage drop can be described as

$$V_{drop} = 81m^2 a^3 \frac{\rho_{supply} L_{max}}{d_{supply} W_{supply} \eta}. \tag{6.21}$$

where m is the number of pixels along the horizontal direction, a the pixel pitch, ρ_{supply} the resistivity of supply line, L_{max} the maximum luminance of display, d_{supply} the thickness of supply line, W_{supply} the width of supply line, and η the current efficiency of the OLED diode.

If the supply line width is proportional to the display size, then

$$W_{supply} = a W_0 \tag{6.22}$$

and

$$V_{drop} = 81m^2 a^2 \frac{\rho_{supply} L_{max}}{d_{supply} W_0 \eta} \tag{6.23}$$

When display resolution is constant, pixel pitch is proportional to the display size; therefore, Eq. (6.23) indicates that voltage drop is proportional to the square of the display size.

Therefore, to achieve a voltage drop equivalent to a 20-in. display, a 30-in. display needs to reduce the $\rho_{supply}/d_{supply}\eta$ value by $\left(\frac{20}{30}\right)^2 = \frac{1}{2.25}$ times; a 40-in. display by $\left(\frac{20}{40}\right)^2 = \frac{1}{4}$ times; and a 50-in. display by $\left(\frac{20}{50}\right)^2 = \frac{1}{6.25}$ times.

To improve the $\rho_{supply}/d_{supply}\eta$ value, tandem OLED technology (discussed in Section 2.6.3) is useful. Figure 6.27 shows the structure of a hypothetical bottom emission device. Using an N-unit tandem OLED device, the operating voltage drop and electric current stress are reduced to $1/N$, which results in N^n times the lifetime. ($n = 1.2–1.9$ as discussed in Section 2.5. If $n = 1.5$ is assumed, two-stack tandem OLED gives 2.82 times, and three-stack gives 5.20 times the lifetime, theoretically.) For example, a white emission + color filter device, discussed in Chapter 7, might use two units (three units in exceptional cases) from this perspective.

When a multiunit stack is used, voltage applied to the device is high. Driver IC cost is increased rapidly as the "design rule" (minimum width in the semiconductor design) is increased, so the operating voltage of device formulation should be minimized by optimization even with the tandem OLED structure case.

For larger displays, wiring with lower resistivity is necessary. Copper can have low resistivity values $\rho_{supply} \sim 2 \times 10^{-6}\Omega \cdot cm$, so $\rho_{supply}/d_{supply}\eta$ can be less than half that of aluminum wiring. Combined with tandem structure, very large displays would be possible using these parameters.

As the minimum requirement of any OLED display component is only a TFT glass substrate and a driver IC, an OLED display will be sufficiently

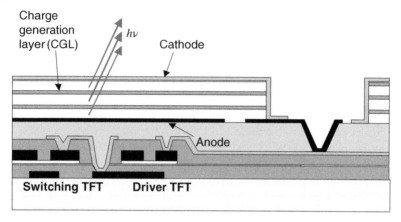

Figure 6.27 Schematic example of an active-matrix-driven tandem OLED structure.

cost-competitive with a large LCD in the near future if the manufacturing infrastructure is sufficiently developed.

6.4.4 LTPS-TFT Pixel Compensation Circuit

The liquid crystal in LCD displays is driven by voltage. (In fact, in the very early stages of LCD development, liquid crystal, e.g., that was used in dynamic scattering-mode LCD, was driven by electric current rather than voltage.) On the other hand, OLED displays are driven by electric current, so small current variations can be distinguished as nonuniformity (*mura*). Thus, a simple driving method using a 2T1C circuit (Figure 4.33), for example, tends to cause yield loss due to mura.

Figure 6.28 shows an example of luminance mura. The human eye is very sensitive to changes in luminance, so if there is driver TFT current nonuniformity in the display area, it is perceived as mura. In particular, LTPS-TFT treated by ELA has a serious mura issue, so various compensation circuits have been proposed and applied in practice.

6.4.4.1 Voltage Programming

The basic principle of V_{th} compensation is to store the driver TFT V_{th} in the gate node as $V_G = V_D + V_{th}$, by connecting gate and drain nodes as shown in Figure 6.29. This principle has been used in various compensation circuits, not only for voltage programming.

Figure 6.30 shows a V_{th} compensation circuit by voltage programming presented by Dawson et al. [9]. This circuit is thought to be the prototype of most subsequent voltage programming compensation circuits [10–15].

Figure 6.28 Image of luminance mura (nonuniformity) due to variation in threshold voltage.

Figure 6.29 Schematic representation of threshold voltage detection via gate–drain node connection.

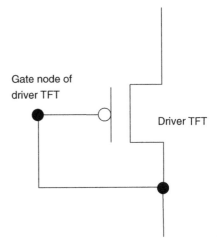

Gate node of driver TFT

Driver TFT

Here, we discuss the working principle of the circuit, step by step:

1. The *AZ* line is set to negative voltage, and MN3 TFT turns on. (Note that the TFT is P-channel.)
2. The gate node and drain node of MN2 are connected and create the same scenario as shown in Figure 6.29, so the gate node of driver TFT (MN2) stores V_{th} information.
3. The data line signal voltage is added to the driver TFT gate node using capacitance coupling of C1.

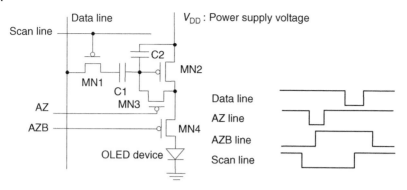

Figure 6.30 Voltage programming circuit drawn by Dawson et al. [9].

4. As the gate node of the driver TFT stores the data line signal plus V_{th}, the source–drain current of the driver TFT does not contain V_{th} information. Therefore, mura (nonuniformity) related to V_{th} variation does not take place, due to the cancellation mechanism.

Voltage programming is the method most frequently used for actual OLED display products. The major reason is that an OLED driver IC is similar to an LCD display driver IC. As many circuit design techniques for LCD drivers can be applied for voltage programming drivers, IC drivers can be designed very easily and are also relatively inexpensive.

On the other hand, IC driver voltage programming has some disadvantages such as the following:

- Inability to compensate for either mobility variation or OLED variation
- Aperture ratio loss due to the occupied space by the programming circuit in the case of bottom emission structure. (Top emission can circumvent this issue.)

Note that the lifetime is proportional to $1/J^n$, as discussed in Section 2.5 (J = current density, $n = 1.2–1.9$). As current density is inversely proportional to the aperture ratio, therefore,

$$\text{Lifetime} \propto \frac{1}{J^n} \propto AR^n$$

If $n = 1.5$, the lifetime becomes only 35% for $AR = 50\%$, 25% for $AR = 40\%$, and 16% for $AR = 30\%$; this is a significant reduction, so a top emission structure is useful to avoid such problems.

6.4.4.2 Current Programming

Figure 6.31 shows one of the most widely known current programming compensations, current mirror pixel circuit designs [16]. The circuit uses four TFTs and one capacitor.

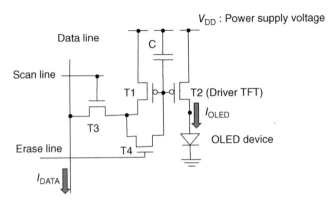

Figure 6.31 Current mirror-type compensation circuit proposed by Sasaoka et al. [16].

Here we assume that the TFT dimension T1 is equal to T2. (In actual implementations, T1 is intentionally designed with a dimension different from that of T2 to reduce the charging time.) Turning on T3 and T4, and also applying current I_{data} to T1, ensures that the gate nodes of T1 and T2 are charged to the target voltage so that the driver TFT can flow current equivalent to I_{data}. This method uses current, not voltage, as a reference, so it can compensate for V_{th} and mobility of the driver TFT and also for any variation in OLED device characteristics.

Although current programming has various advantages, it has drawbacks such as driver IC cost (discussed above), and also its long charging period for low grayscale levels. Several methods have been proposed to address this issue.

6.4.4.3 External Compensation Method

There is minimal change in LTPS-TFT performance during the driving operation. Therefore, mura (nonuniformity) can be corrected by applying an offset to the pixel by the driver IC using a factory-set value. (This is called the *external compensation* method[17, 18].) This method is promising because the cost of memory is steadily decreasing and also necessary frame buffer to store the mura information can be minimized due to the recent evolution of data compression algorithms. The information necessary to correct mura can be obtained by either electrical or optical measurements. Figure 6.32 presents an example of a product using the external compensation method by electrical inspection during the fabrication, referred to as global mura compensation (GMC). Actual uniformity data before and after implementation of the GMC method are shown in Figure 6.33.

Figure 6.34 shows the process flow of the GMC external compensation method.

The manufactured panels are inspected pixel by pixel (actual inspection equipment is shown in Figure 6.35), and the offset data are recorded in

Figure 6.32 A portable television set using the external global mura compensation method (GMC) (by Eastman Kodak).

the memory mounted on each panel. Luminance error (mura) caused by LTPS-TFT current variation can be canceled by calculating the correct driver TFT gate node value (R′, G′, and B′) using an offset measured in the factory when displaying R, G, and B signals on a display.

6.4.4.4 Digital Driving

In most OLED driving methods (analog driving), the TFT is in the saturation region condition, $V_{DS} > V_{GS} - V_{th}$ (where V_{DS} is drain voltage relative to source voltage and V_{GS} is gate voltage relative to source voltage). This is because linear region driving ($V_{DS} < V_{GS} - V_{th}$) causes wide variations in TFT performance. On the other hand, saturation region driving consumes more power than does linear region operation. Normal analog driving applies almost the same source–drain voltage for both OLED and TFT drain. If the linear region method can be adopted, power consumption can be reduced by almost 50%.

To obtain this advantage, digital driving [19, 20] employs time frame modulation; specifically, the driver IC turns the current flow on and off, rather than changing the OLED current by controlling the gate node voltage of the driving TFT.

Digital driving resembles plasma display driving in terms of rapid ON/OFF switching, so it has similar issues, such as pseudocontour.

As shown in the upper segment of Figure 6.36, digital driving turns on and off multiple subframes in one frame period. (For a 60-Hz display, $\frac{1}{60} = 16.7$ is one frame period.)

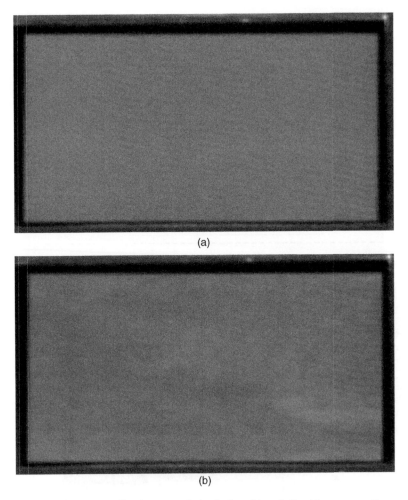

Figure 6.33 Images of luminance variation before (b) and after (a) compensation.

Here, we assume that the Nth line is ON (white) and $(N + 1)$th line is OFF (black) during subframe periods 1–4. We also assume that the Nth line is OFF and the $(N + 1)$th line is ON during the fifth subframe. In this condition, if the line of sight is moved from the Nth line to $(N + 1)$th line, the line of sight remains in the ON region, so a white line is observed. On the other hand, if the line of sight is moved from the $(N + 1)$th line to the Nth line, the line of sight line remains in the OFF region, so a black line is observed. This is the crux of the pseudocontour.

Pseudocontour has been effectively addressed in plasma display technology, and there are also many effective driving methods for avoiding pseudocontour

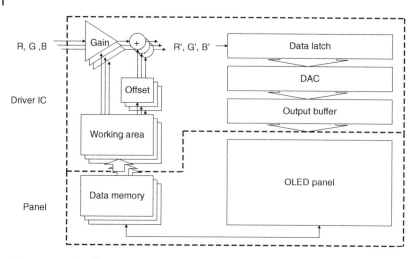

Figure 6.34 Flowchart showing operation of external GMC method.

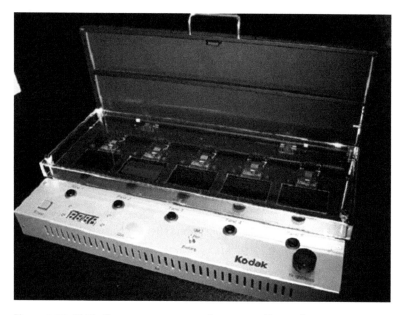

Figure 6.35 GMC offset measurement equipment used in production.

in OLED displays. In the case of a normal digital driving scheme, a subframe is selected according to a binary number system. In binary number system emission, a consecutive grayscale can have a very different emission timing, so pseudocontour is prominent in such a case because the discontinuous emission occurs when the eye is moved. To circumvent this issue, (1) rearrangement

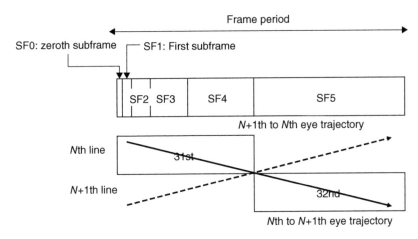

Figure 6.36 Schematic representation of the pseudocontour mechanism.

of subframe and (2) distribution of high-intensity emission in a frame by the error-diffusion method are used so that the pseudocontour becomes less visible to the human eye.

Digital driving has another issue related to the lifetime. In analog driving, resistance of a driver TFT in a pixel is comparable to that of the OLED device, as the driver TFT is driven in saturation region. So when the OLED resistance is increased due to device degradation, the change in voltage drop across the TFT-OLED connection is not very large. However, in the case of digital driving, driver TFT is driven in linear region, so the TFT resistance is much lower than that of analog driving. As a result, TFT-OLED connection voltage drop is much more subjected to the resistance change of the OLED device, so the lifetime of pixel becomes shorter in the case of digital driving. To circumvent this situation, compensation of OLED device degradation has been proposed [21].

As discussed, digital driving has large merit in power reduction. However, due to recent trends toward high resolution, the switching speed is not sufficient for such high resolution, so the actual product implementation of digital driving is decreasing.

6.4.5 Circuit Integration by LTPS-TFT

As LTPS-TFT has a high driving capability, it is possible to form various circuits on substrates, such as shift registers or multiplexers. Figure 6.37 shows an example of an integrated circuit.

A *data driver* circuit is composed of video signals 1–3 (red, green, and blue); a sampling circuit, which delivers video signals to the data line; and a shift register, which controls the timing of the sampling circuit.

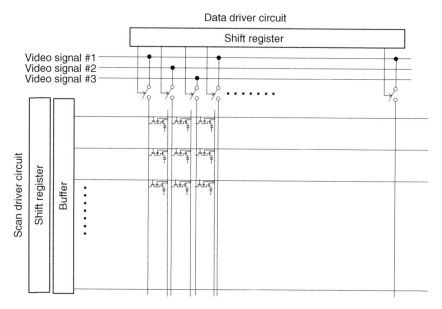

Figure 6.37 Schematic representation of circuit integration using low-temperature polycrystalline silicon (LTPS) TFTs.

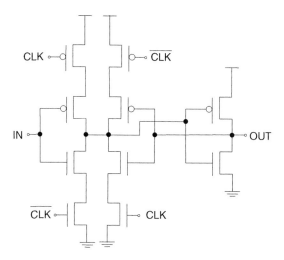

Figure 6.38 Schematic representation of a CMOS-type static shift register.

A *scan driver* delivers signal to the selected scan line by the shift register via a buffer circuit.

As an example of circuit integration, we shall discuss the shift register circuit. For a shift register, a circuit such as that shown in Figure 6.38 is used for CMOS formats, and circuits such as those shown in Figures 6.39 and 6.40 are for NMOS and PMOS formats.

Although CMOS technology involves many process steps (discussed in Section 6.2.1) and is complicated, with multiple TFTs in the same circuitry, it is capable of driving at speeds exceeding 10 MHz. It also has relatively low power consumption and high reliability.

On the other hand, NMOS and PMOS single-channel technologies can employ short process steps, which is advantageous in terms of cost (see

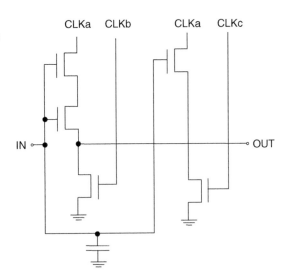

Figure 6.39 Schematic representation of an NMOS-type dynamic shift register.

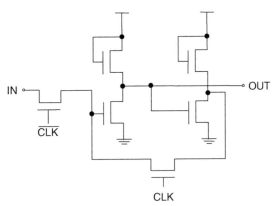

Figure 6.40 Schematic representation of an NMOS-type static shift register.

Section 6.2.1). However, the driving speed is not very high. For example, the dynamic shift register shown in Figure 6.39 has a driving speed of 2 MHz, and the static shift register shown in Figure 6.40 has a driving speed of 500 kHz.

6.5 TFT TECHNOLOGIES FOR OLED DISPLAYS

There are many difficulties with the LTPS-TFT formation process using ELA, such as laser stability, TFT performance variation, poor laser process productivity, and lack of substrate size scalability. Here we discuss new approaches to these issues.

6.5.1 Selective Annealing Method

Conventional ELA method irradiates laser over whole substrate by step-and-repeat manner, as discussed in Section 6.2.4. However, the area to be crystallized is only TFT channels, so whole-area emission wastes laser power, which reduces the manufacturability. Several methods have been developed to irradiate the laser light only to desired locations.

6.5.1.1 Sequential Lateral Solidification (SLS) Method

The SLS method [22] employs a technique that causes silicon grains to grow perpendicularly to the boundary between the liquefied region (in which laser melting occurs) and the solid region. With this method, the grain sizes that can be obtained are about two- to threefold larger than those obtained by conventional ELA.

The SLS method controls the exposed and unexposed regions of the substrate by means of a mask that has a long line-shaped aperture. Figure 6.41 shows an example of the so-called two-shot SLS method, in which, with the first excimer laser pulse, every alternate location is molten. Crystallization takes place using a precrystallized location as a crystallization seed, so that a long crystal grain can be obtained. Using a narrow laser irradiation width (e.g., 5 μm), any spurious crystallization from the seed beyond the grain boundary can be suppressed, thus ensuring a high-quality large grain. In the SLS method, the energy profile of the beam is almost rectangular, so virtually no energy gradient occurs; thus, no overlap is required, unlike the case with conventional ELA, and this is a major advantage in terms of manufacturing throughput.

This method has been used for many of the small- to-medium-size active-matrix OLED displays on the market.

6.5.1.2 Selective Annealing by Microlens Array

Sugimoto et al. proposed a method to irradiate laser beam using microlens array only to the TFT location using Nd : YAG pulsed laser [23] (Figure 6.42).

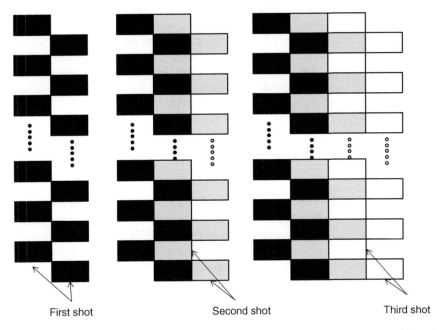

First shot   Second shot   Third shot

Figure 6.41 Schematic example of a display obtained using the two-shot sequential lateral solidification (SLS) method.

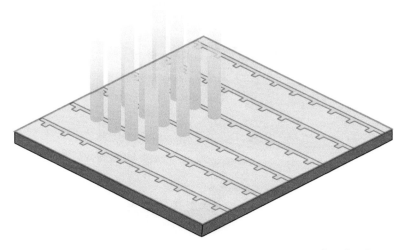

Figure 6.42 Schematic example of the selective annealing process by micro lens array.

The system also employs a real-time mask alignment system, which enables the system to be applied to very large substrates.

Utsugi et al. presented that Sakai Display Products successfully fabricated LTPS-TFT LCD panel using this method. (PLAS: Partial Laser Anneal Silicon technology) [24]. In the past, due to the size restriction of excimer laser process, largest mother substrate used for LTPS was limited up to generation 6 (1500 mm × 1850 mm) so far when generation 10.5 (2940 mm × 3370 mm) amorphous silicon TFT plant is under construction. The selective annealing method has eliminated one important bottleneck for LTPS to use large substrates.

6.5.2 Microcrystalline and Superamorphous Silicon

The mobility of LTPS-TFT formed by ELA OLED displays fluctuates with various factors such as frequent gas exchange, laser energy instability, and transmittance loss of laser optics due to UV light and electrode degradation, so process maintenance may be problematic with the use of this method. Also, it may be difficult to apply to large substrates because of the laser width restriction due to the laser power limitation.

In 2003, Tsujimura et al. fabricated a 20-in. superamorphous silicon OLED display prototype and proved that large OLED displays could be produced without using excimer laser annealed LTPS-TFT (Figure 6.43) [25].

Figure 6.43 A large OLED display prototype obtained using a non-LTPS substrate (demonstration by IDTech/IBM/CMO in 2003).

With the parameters $\mu = 0.5$ cm^2/(V·s) (mobility of amorphous silicon) and $L = 5$ μm, the necessary TFT channel width should be in the range of hundreds of micrometers, which is larger than the pixel size. It would be possible to incorporate a large TFT in a pixel by zigzag formation, but the defect number might increase due to the long electrodes formed very closely. For this reason, Tsujimura et al. introduced the possibility of reducing the TFT dimensions through mobility improvement by using modified a-Si and introducing microcrystalline Si [26, 27]. It is believed that normal a-Si has too much instability to cope with, so more stable bonding would be necessary. If both high mobility and long-term reliability can be realized, this would represent a major breakthrough in attempts to achieve large AMOLED displays, so it was attempted to improve both Si bonding by low-cost crystallization [28]. Circuit design improvements, such as developing a pixel-level compensation circuit, are also widely discussed to cope with the instability [29, 30]. Instability of a-Si and microcrystalline silicon is mainly due to the threshold voltage shift, so voltage programming discussed in Section 6.4.4 can be applied.

The graph in Figure 6.44 indicates the process condition under which either amorphous silicon (a-Si) or microcrystalline silicon (μc-Si) can be obtained by $SiH_4 + H_2$ gas processing using the PECVD method. As the introduced power is increased, the crystallization ratio improves. Depending on the conditions,

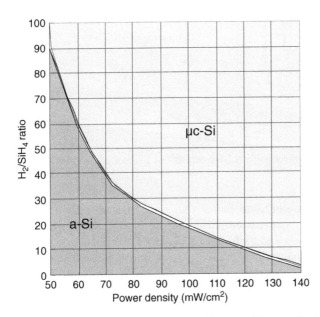

Figure 6.44 Graphical depiction of crystalline condition monitoring using the plasma-enhanced chemical vapor deposition (PECVD) method.

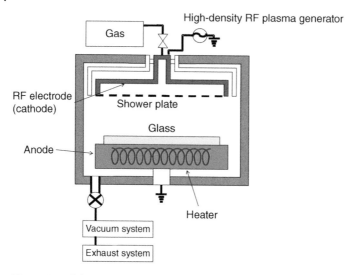

Figure 6.45 Schematic illustration of a high-density plasma CVD apparatus.

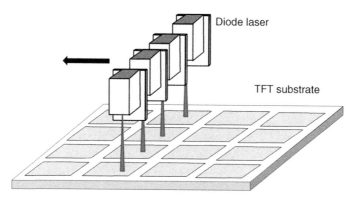

Figure 6.46 Diagram illustrating formation of microcrystalline silicon induced by diode laser annealing.

μc-Si can have higher mobility and better reliability than a-Si (amorphous silicon). Several methods can be employed to form μc-Si, including the layer-by-layer (LBL) method and high-density plasma CVD (Figure 6.45). For such methods, conventional a-Si manufacturing equipment can be used, so it is advantageous from an investment perspective and is also cost-effective because the process step is the same as that in a-Si TFT formation.

Figure 6.46 illustrates equipment used for μc-Si formation by the *diode laser thermal annealing* (DLTA) method, which converts amorphous silicon to microcrystalline silicon by diode laser annealing. In the case of microcrystalline silicon formation by PECVD, the process employs deposition and etching simultaneously, so it tends to result in poor thickness distribution and low deposition rate. The DLTA method can be used to avoid such problems, but the number of process steps might increase due to the use of laser annealing and etching-stopper type TFT.

The approach to use amorphous silicon and microcrystalline silicon were used for only several products. However, it was using NMOS TFT, so many development results of amorphous silicon TFT driving, such as NMOS + common cathode device structure in Figure 4.37, anode line structure in Figure 4.38, double-channel structure and pixel compensation circuit were applied to oxide semiconductor (discussed in Section 6.5.4) driving, which became mainstream of OLED television later on.

6.5.3   Solid-Phase Crystallization

Other methods can be employed to achieve the performance between amorphous silicon and laser-annealed polycrystalline silicon; the most popular methods are metal-induced crystallization (MIC) and alternating magnetic field crystallization (AMFC).

6.5.3.1   MIC and MILC Methods

The MIC method [31] applies the theory that the crystallization temperature of silicon in contact with nickel—which has the same body-centered structure as does silicon crystal—is reduced. The MIC method is widely used to control the grain boundary location by controlling the location of Ni seed to ensure crystallization growth in the lateral direction as shown in Figure 6.47. This method to control the grain boundary location by lateral crystalline growth is called *metal-induced lateral crystallization* (MILC). Figure 6.47a shows how crystalline growth takes place from both ends of the channel and then the crystals collide with each other in the center when nickel is deposited on the TFT and annealed. Processes for controlling the location of silicon grain, as in this MILC method, are collectively referred to as *location-controlled polysilicon*. Figure 6.47b shows a process in which the gate insulator aperture is opened to allow the crystallization to propagate from the area in contact with the aperture.

The MILC method does not require laser annealing control; however, the presence of any residual nickel introduces the risk of current leakage and, thus, impaired reliability. Studies are underway to reduce the amount of residual nickel.

6.5.3.2   AMFC Method

The main objective of the alternating magnetic field crystallization (AMFC) method is to reduce the crystallization temperature by the addition of an alternating magnetic field [32, 33].

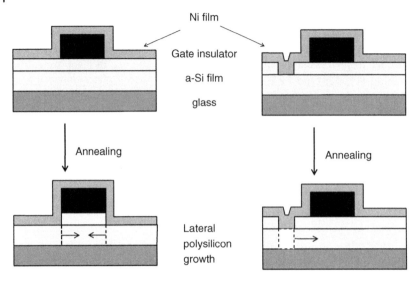

Figure 6.47 Schematic illustration of the principle of solid-phase crystallization using the metal-induced lateral crystallization (MILC) method.

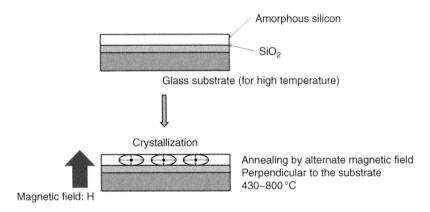

Figure 6.48 Diagram of crystallization apparatus employing the alternating magnetic field crystallization (AMFC) method.

Figure 6.48 shows crystallization using the AMFC method described by Seo et al. [34]. After deposition of SiO_2 by PECVD, 500 Å of amorphous silicon is deposited using $SiH_4 + H_2$ gas, and the amorphous silicon film is crystallized by a 500 °G (230 kHz, 50 A, 8 s) alternating magnetic field at ~430–550 °C, followed by 5 min of annealing.

The AMFC method can cause crystallization without the concern of the nickel-related issues that usually occur with the MIC or MILC method.

6.5.4 Oxide Semiconductors

Oxide semiconductor TFTs [35–38] such as those made from ZnO or InGaZnO (also abbreviated IGZO) are attracting much attention for several reasons:

- The manufacturing process for oxide semiconductors is almost the same as that for amorphous silicon TFTs; therefore, it has cost merit compared with LTPS-TFT by ELA.
- Much higher mobility can be achieved for oxide semiconductors than for amorphous silicon, which enables integration of higher-speed circuits for low circuit cost.
- The manufacturing process is very similar to amorphous silicon TFT, so same manufacturing line can be used by adding minimum investment, such as, only oxide semiconductor deposition equipment, which promises low manufacturing cost.
- There is less performance variation related to driving stress than with amorphous silicon.
- The process temperature is lower, which is suitable for plastic substrates.
- As the channel layer is transparent, transparent displays may be possible.

The key reason of high performance with relatively easy processing lies in its conduction mechanism. The conduction band minimum (CBM) of oxide semiconductors are formed by s orbital of metal atoms. Insufficient oxygen brings about the free electron, so it normally shows n-type carrier conduction. The bandgap is more than 3 eV, which causes transparent appearance regarding the visible light wavelength range.

It is known that silicon-based carrier conduction is governed by sp^3 hybrid conduction mechanism as shown in Figure 6.49. For maximum conduction, large overlap of sp^3 orbital is necessary. However, the orbital shape has much anisotropy, so a little disorder causes large loss of overlap that reduces the semiconductor mobility. As a result, for example, mobility of polycrystalline TFT and amorphous silicon TFT has more than two orders of magnitude difference in field-effect mobility.

In n-type oxide semiconductors, the conduction is based on s orbital overlap as shown in Figure 6.50, which has spherical orbital shape. Spherical orbital promises good conduction even when some disorder takes place [39].

Density of states is very low in the gap, so Fermi level easily exceeds the CBM and shows the carrier transport based on the band conduction mechanism, similar to that of crystalline semiconductors, different from amorphous silicon whose driving region is in the tail state. As it is N-channel conduction with wide band gap, the hole density is very low, so no inversion region can be made, which promises very low leakage current in the negatively biased operation.

For best performance during manufacturing, many aspects must be taken into account [40].

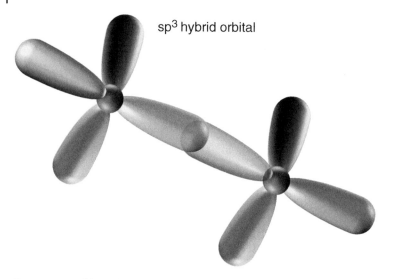

sp³ hybrid orbital

Figure 6.49 sp³ hybrid orbital conduction mechanism by silicon-based semiconductors.

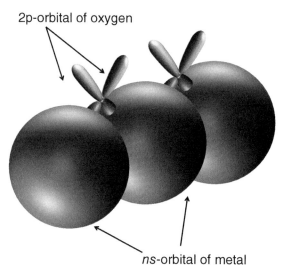

2p-orbital of oxygen

ns-orbital of metal

Figure 6.50 s-orbital-based conduction mechanism by oxide semiconductors.

- High mobility
- Uniform and stable amorphous film deposition
- Low-temperature deposition, compatible with low T_g substrate, such as plastic. (Currently, it is known that oxide semiconductor needs high-temperature annealing for stable operation.)
- Low carrier concentration deposition.

For high mobility, a heavy atom ion, such as In or Sn, which has large expanded ns ($n > 5$) orbital as its least unoccupied state, will be effective (e.g., ITO or SnO_2). However, the demerit of such systems is that it easily causes oxygen vacancy and uncontrollable high density carrier is introduced.

To circumvent the situation, controllability of carrier density is important. By adding Ga, the oxygen vacancy can be controlled. Thus, $In_3-ZnO-Ga_2O_3$ system, so-called IGZO was introduced [41]. There is also an approach to use other type of crystalline oxide semiconductor systems [42].

Though oxide semiconductors, such as IGZO, show much better stability performance, they have disadvantage under illuminated condition. In the case of IGZO, occupied state higher than the CBM absorbs the short wavelength light, and threshold voltage of TFT is negatively shifted by hole creation. The phenomenon is accelerated especially in negatively biased condition. (Negative illumination bias stress: NIBS.)

Oxide semiconductors are normally deposited by sputtering or the atomic layer deposition (ALD) method. The sputtering method has been widely used for metal wiring or incorporation of pixel electrodes in TFTs and thus has achieved mature technology status in the industry. On the other hand, the ALD method can achieve very high-quality deposition, so it has the advantages of mobility and reliability. For the manufacturing of displays, sputtering is popularly used.

In the TFT fabrication process, defects are created, which causes the trap centers. To assure the stability and uniformity, thermal annealing over 350 °C is often used. Also, it is important to consider the effect of hydrogen, which can introduce oxygen vacancy in the oxide semiconductor layer.

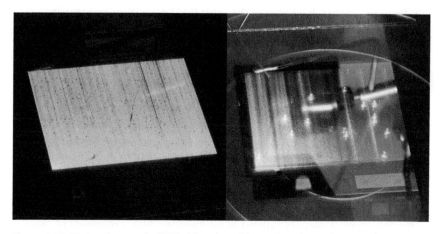

Figure 6.51 First active-matrix OLED driven by oxide semiconductor (courtesy by S.H. Park).

Figure 6.52 OLED television product using oxide semiconductor TFT.

Oxide semiconductor is N-channel, so the pixel circuit, driving, and TFT structure are similar to the case of amorphous silicon and micro-crystalline TFT, discussed in Section 6.5.2. Same approaches such as common-cathode + bottom emission (Figure 4.37), anode-line structure (Figure 4.38), double-channel, and compensation circuit have been applied.

Figure 6.51 shows the first active-matrix OLED display driven by oxide semiconductor [43, 44]. In this prototype, both TFT and OLED devices are transparent so that the transparent feature of oxide semiconductor can be fully featured. Figure 6.52 shows the AMOLED television product by oxide semiconductor. Due to the scalability of technology, oxide semiconductor has acquired the state-of-the-art position of OLED television.

References

1 J. Kanicki, *Amorphous and Microcrystalline Semiconductor Devices: Materials and Device Physics*, Artech House Materials Science Library, 1992.
2 E. Chen, "Fabrication Techniques III: Etching", IRG 3: Interface-Mediated Assembly of Soft Materials, pp. 1-18, Harvard School of Engineering and Applied Science (2016).
3 I. Sumita, T. Nagata, A. Gonjyo, T. Sado, and M. Miyao, Bonding fluctuation effect of non-crystallized Si on dielectric film, *Proceeding of 48th meeting by Japan Society of Applied Physics*, 31a P16-3 (2001).

4 G. Rajeswaran et al., Active matrix low temperature poly-Si TFT/OLED full color displays: development status, *SID 2000 Proc.*, p. 974 (2000).

5 T. Tsujimura, W. Zhu, S. Mizukoshi, N. Mori, K. Kawabe, M. Kohno, and K. Onomura, Design for the OLED as the best display technology, *OLEDs Asia 2006 Proc.*, (2006).

6 S. M. Sze and K. K. Ng, *Physics of Semiconductor Devices*, Wiley-Interscience, New York, 1969.

7 S. Zhang et al., Reduction of off-current in self-aligned double-gate TFT with mask-free symmetric LDD, *IEEE Trans. Electron Devices* **48**:1490–1494 (2002).

8 T. Kawakita et al., Analysis of hot carrier effect in low temperature poly-Si gate-overlapped lightly doped drain thin film transistors, *Jpn. J. Appl. Phys.* **42**:3354 (2003).

9 R. M. A. Dawson et al., Design of an improved pixel for poly-silicon active-matrix organic LED display, *SID 1998 Digest*, pp. 11–14 (1998).

10 S. Ono, Y. Kobayashi, K. Miwa, and T. Tsujimura, Pixel circuit for a-Si AM-OLED, *Proceedings of the International Display Workshop 2003*, p. 255 (2003).

11 D. Fish, N. Young, M. Childs, W. Steer, D. George, D. McCulloch, S. Godfrey, M. Trainer, M. Johnson, A. Giraldo, H. Lifka, and I. Hunter, A comparison of pixel circuits for active matrix polymer/organic LED displays, *SID 2002 Digest*, p. 968 (2002).

12 S. Ono, K. Miwa, Y. Maekawa, and T. Tsujimura, Shared pixel compensation circuit for OLED displays, *Proceedings of the 9th Asian Symposium on Information Display* (2006).

13 J. L. Sanford and F. R. Libsch, Vth compensation performance of voltage data AMOLED pixel circuits, *IDRC 2003 Digest*, p. 38 (2003).

14 N. C. van der Vaart, E. A. Meulenkamp, N. D. Young, and M. Fleuster, Next-generation active-matrix polymer OLED displays, *Asia Display/IMID 04 Proc.* (2004).

15 S. Ono, K. Miwa, Y. Maekawa, and T. Tsujimura, Vth compensation circuit for AMOLED displays composed of two TFTs and one capacitor, *IEEE Trans. Electron Devices* **54**(3), pp. 462–467 (2007).

16 T. Sasaoka et al., A 13.0-inch AM-OLED display with top emitting, *SID 2001 Proc.*, p. 384 (2001).

17 T. Tsujimura, W. Zhu, S. Mizukoshi, N. Mori, M. Yamaguchi, K. Miwa, S. Ono, Y. Maekawa, K. Kawabe, M. Kohno, and K. Onomura, Advancements and outlook of high performance active-matrix OLED displays, *SID 2007 Digest*, p. 84 (2007).

18 S. Mizukoshi, Driving circuit for OLED display and its technology trend, *FPD Intnatl. Technology Forum Proc. E-5* (2007).

19 K. Kawabe, T. Tsujimura, S. Mizukoshi, N. Mori, K. Onomura, S. VanSlyke, J. P. Spindler, A. Pleten, and K. M. Fallon, Digital drive with white OLEDs, *SID 2009 Digest*, p. 5 (2009).

20 H. Akimoto et al., An innovative pixel-driving scheme for 64-level gray-scale full-color active matrix OLED displays, *SID 2002 Digest*, pp. 972–975 (2002).

21 K. Oh, S. Hong, and O. Kwon, Lifetime extension method for active matrix organic light-emitting diode displays using a modified stretched exponential decay model, *IEEE Electron Device Lett.* **36**(3) (2015).

22 J. B. Choi et al., Sequential lateral solidification (SLS) process for large area AMOLED, *SID 2008 Digest*, p. 97 (2008).

23 S. Sugimoto, T. Kiguchi, M. Hatanaka, M. Mizumura, K. Kajiyama, J. Kido, Selective laser-annealing system for LTPS-TFT panels, *SID Symposium Digest of Technical Papers*, Vol. **46**, Issue 1, pp. 394-397 (2015).

24 S. Utsugi, N. Nodera, T. Matsumoto, K. Kobayashi, and T. Oketani, Novel LTPS technology for large substrate, *SID Symposium Digest of Technical Papers*, Vol. **47**, Issue 1, pp. 915-918 (2016).

25 T. Tsujimura et al., A 20-inch OLED display driven by super-amorphous-silicon technology, *SID 2003 Proc.*, p. 6 (2003).

26 T. Tsujimura, F. Libsch, and P. Andry, Amorphous silicon thin film transistor for large size OLED television driving, *J. SID* **13**(2):161 (2005).

27 T. Tsujimura, Amorphous/microcrystalline silicon thin film transistor characteristics for large size Oled television driving, *Jpn. J. Appl. Phys.* **43**(8A):5122–5128 (2004).

28 K. Girota et al., A 14-inch uniform AMOLED display with low cost PECVD based microcrystalline silicon TFT backplanes, *SID 2008 Digest*, p. 1289 (2008).

29 Y. He, R. Hattori, and J. Kanicki, Improved A-Si:H TFT pixel electrode circuits for active-matrix organic light emitting displays, *IEEE Trans. Electron Devices* **48**(7):1322 (2001).

30 T. Hasumi, S. Takasugi, K. Kanoh, and Y. Kobayashi, New OLED pixel circuit and driving method to suppress threshold voltage shift of a-Si:H TFT, *SID Digest*, pp. 1547–1550 (2006).

31 S. Y. Yoon, K. H. Kim, C. O. Kim, J. Y. Oh, and J. Jang, Low temperature metal induced crystallization of amorphous silicon using Ni solution, *J. Appl. Phys.* **82**(2):5865–5867 (1997).

32 S.-K. Hong, B.-K. Kim, and Y.-M. Ha, LTPS technology for improving the uniformity of AMOLEDs, *SID 2007 Digest*, p. 1366 (2007).

33 S. H. Jung, H. K. Lee, C. Y. Kim, S. Y. Yoon, C. D. Kim, and I. B. Kang, 15-inch AMOLED display with SPC TFTs and a symmetric driving method, *SID 2008 Digest*, p. 101 (2008).

34 H. S. Seo, C. D. Kim, I. B. Kang, I. J. Chung, M. C. Jeong, J. M. Myoung, and D. H. Shin, Alternating magnetic field-assisted crystallization of Si films without metal catalyst, *J. Crystal Growth* **310**(24):5317–5320 (2008).

35 H. Hosono and Y. Abe, Porous glass-ceramics composed of a titanium phosphate crystal skeleton: a review, *J. Non-Cryst. Solids* **190**(3):185–197 (1995).

36 S. H. K. Park, M. Ryu, S. Yang, C. Byun, C.-S. Hwang, K. I. Cho, W.-B. Im, Y.-E. Kim, T.-S. Kim, Y.-B. Ha, and K.-B. Kim, Oxide TFT driving transparent AM-OLED, *SID 2010 Digest*, p. 245 (2010).

37 J. K. Jeong et al., 12.1-inch WXGA AMOLED display driven by indium-gallium-zinc oxide TFTs array, *SID 2008 Digest*, p. 1 (2008).

38 E. Fukumoto, T. Arai, N. Morosawa, K. Tokunaga, Y. Terai, T. Fujimori, and T. Sasaoka, High mobility oxide semiconductor TFT for circuit integration of AM-OLED, *IDW 2010 Proc.*, p. 631 (2010).

39 T. Kamiya, *Transparent Oxides as Active Electronic Materials and Their Applications*, CMC Press, p. 30 (2006).

40 K. Nomura, *Transparent Oxides as Active Electronic Materials and Their Applications*, CMC Press, p. 103 (2006).

41 M. Orita, H. Ohta, M. Hirano, S. Narushima, and H. Hosono, Amorphous transparent conductive oxide $InGaO_3(ZnO)m$ ($m \leq 4$): a Zn4s conductor, *Phil. Mag. B* **81**:501 (2001).

42 S. Yamazaki, Research, development, and application of crystalline oxide semiconductor, *SID2012 Digest*, Vol. **43**, Issue 1, pp. 183-186 (2012).

43 S. H. Park, C. S. Hwang, J. Lee, S. M. Chung, Y. S. Yang, L. Do, and H. Y. Chu, Transparent ZnO thin film transistor array for the application of transparent AM-OLED display, *Issue SID Symposium Digest of Technical Papers*, Vol. **37**, Issue 1, pp. 25-28, (2006).

44 S. H. Park, C. S. Hwant, M. Ryu, S. Yang, C. Byun, J. Shin, J. Lee, K. Lee, M. S. Oh, and S. Im, Transparent and photo-stable ZnO thin-film transistors to drive an active matrix organic-lightemitting-diode display panel, *Adv. Mater.* **21**:678–682 (2009).

7

OLED Television Applications

The success of OLED technology in television applications is due to the superiority of its parameters, such as response speed, viewing angle, and contrast ratio.

In this section, we discuss the factors and methods involved in the design and manufacture of medium-to-large OLED displays and how to achieve high productivity, low cost, and high production yield to compete with other display technologies in the television industry.

7.1 PERFORMANCE TARGET

Table 7.1 shows the specification of commercially available OLED television [1]. As OLED is self-emissive device, there is no light emission in the black state. Therefore, the contrast ratio is infinite in dark rooms, which gives impressive image to the viewers. Also large color gamut can give fidelity to the intended image signals.

OLED power consumption depends on the image, as can be seen in the table. As the luminance is increased, the power consumption increases. With full-screen black image, the power consumption can be very small, which is mainly consumed by tuner and other electronic components. It has been reported that the average luminance of television is about 21.2% of white screen [2], so the display power consumption can be considerably reduced with normal television broadcasting condition.

OLED has been used mainly for small mobile displays. To enjoy the OLED's superior performance for television, such as excellent contrast, viewing angle, and response time, it is necessary to make large OLED television economically available for purchase by many users. To make that happen, high-yield manufacturing is inevitable.

OLED Display Fundamentals and Applications, Second Edition. Takatoshi Tsujimura.
© 2017 John Wiley & Sons, Inc. Published 2017 by John Wiley & Sons, Inc.

Table 7.1 Specification of Commercially-Available OLED Television (As of September 2016)

	OLED-TV1		OLED-TV2	
Display size	54.6-in.		54.6-in.	
Resolution	1920×1080 progressive		1920×1080 progressive	
Type	Curved		Curved	
Display mode	Dynamic	Standard	Vivid	Standard
Luminance of white (Standard) (cd/m^2): L_w	489	398	373	389
Luminance of black (cd/m^2):	0	0	0	0
Contrast ratio (dark room)	$\infty : 1$	$\infty : 1$	$\infty : 1$	$\infty : 1$
Color gamut (versus sRGB)	116%	120%	118%	118%
CCT(K)	12,580	8,945	11,030	10,230
Power (full-screen white) (W): P_W	252	219	213	210
Power(full-screen black) (W): P_K	115		51	

OLED-TV1: Samsung KN55S9CAFXZA, OLED-TV2: LG55EA9800.

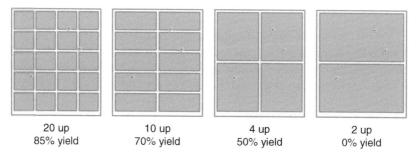

| 20 up | 10 up | 4 up | 2 up |
| 85% yield | 70% yield | 50% yield | 0% yield |

Figure 7.1 Impact of defect to the display production yield.

Figure 7.1 shows the relationship between defect and yield. In the figure, there are three defects assumed. As shown in the most right figure, if we get two large panels from one mother glass substrate, the production yield would be 0%. If we get four displays, the yield would be 50%. If 10 displays, we can get 70% yield and if 20 displays, 85% yield with this assumption. This suggests that it is not very easy to get reasonable production yield for large displays. A strategy to get reasonable production yield for large OLED display manufacturing is discussed in the following section.

7.2 SCALABILITY CONCEPT

7.2.1 Relationship between Defect Density and Production Yield

The production yield, discussed in Figure 7.1, can be estimated using simulation theory. For the large display manufacturing, it is quite important to judge the cost increase caused by the yield loss due to the form factor change.

7.2.1.1 Purpose of Yield Simulation

Production yield simulation has been studied in the semiconductor and LCD display fields to avoid the potential for any risk associated with the production of large devices or high-resolution display [3]. The product yield simulation technique can also be applied to OLED to some extent to estimate the amount of potential risk.

7.2.1.2 Defective Pixel Number Estimation Using the Poisson Equation

The probability of k pieces of defective pixels existing on a display can be expressed as follows (Poisson distribution):

$$P_{SC}\{k\} = \frac{\lambda_{SC}^k}{k!} e^{-\lambda_{SC}} \tag{7.1}$$

Here, λ_{SC} is the average number of defective pixels on a display. If N_{SCT} is defined as the criterion for the maximum allowable number of defective pixels, production yield related to point defect can be expressed as follows:

$$Y_{SC} = \sum_{k=0}^{N_{SCT}} \frac{\lambda_{SC}^k}{k!} e^{-\lambda_{SC}} \tag{7.2}$$

Figure 7.2 compares the defective pixel number data for small and large displays by simulation. Smaller displays tend to have fewer defective pixels, and larger displays tend to have a higher number of defective pixels. In the most extreme case, the number of defective pixels is proportional to the area. Thus, a 32-in. display can have $(32/2)^2 = 256$ times the number of defective pixels in a 2-in. display.

On the other hand, the maximum allowable number of defective pixels (customer specification) does not increase with increase in display size, which causes significant reduction of production yield for large displays.

7.2.2 Scalable Technology

There is a new trend in the OLED industry to ensure the high production yield of large AMOLED displays by using a combination of scalable technologies including the white OLED emission + color filter method [4, 5]. Gen 8

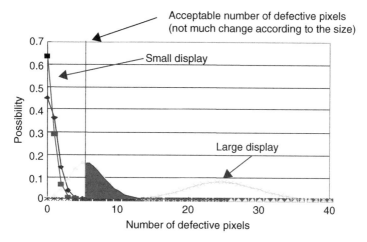

Figure 7.2 Graph illustrating defective pixel number simulation.

investment using this scalable method has been made to achieve low-cost 55-in. OLED television displays [6]. In this section, the scalable technology scenario is explained.

7.2.2.1 Scalability

One of the most important objectives in the manufacture of large TV displays is to achieve low defect density. Table 7.2 indicates which display production techniques are scalable and appropriate for manufacture of large displays.

White + color filter technology, solid-phase crystallization, amorphous silicon, microcrystalline silicon, and oxide TFT do not provide much constraint against enlarging a display, so they can be classified as scalable.

As scalable technologies do not pose much risk in the enlargement of display dimensions, if a display produced by the combination of scalable technologies can provide competitive product-level performance, and if the prototype can

Table 7.2 Examples of Scalable Technology

Technique	Small Display	Large Display
Shadow mask RGB pixilation	OK	Low yield
White OLED + color filter	OK	OK
Polysilicon TFT by ELA	OK	Tool restriction
SPC polysilicon	OK	OK
a-Si, microcrystalline Si	OK	OK
Oxide TFT	OK	OK

attract the customer's interest, it will mean that medium-to-large AMOLED displays can be brought to the market with low risk.

To support this idea, the following three methods have been implemented together successfully to develop a 100% NTSC AMOLED display prototype [4]:

1. The white + color filter method, which can be used to produce a scalable OLED device
2. The *global mura compensation* (GMC) method, which is a scalable driving technology
3. The *solid-phase crystallization* (SPC) method, a scalable TFT backplane method.

Point 1 is discussed in Sections 7.3 and 7.4. Also Points 2 and 3 are discussed in Sections 6.4.4.3 and 6.5.4.1. The prototype detail is described in Section 7.4.1.4.

7.3 MURDOCH'S ALGORITHM TO ACHIEVE LOW POWER AND WIDE COLOR GAMUT

7.3.1 A Method for Achieving Both Low Power and Wide Color Gamut

To create a large display, the white + color filter approach is advantageous in terms of production due to its scalability and the maturity of color filter manufacturing process, which has been used in LCD industry for a long time. In spite of the secured production with high yield, a trade-off between display luminance and color gamut happens, as the color filter absorbs the light coming out from the OLED devices. To make competitive display performance with other display technologies, such trade-offs must be removed. Recent booming OLED televisions employ a method that can ensure a wide color gamut as well as low power consumption. Here we discuss a method that can help provide these.

Figure 7.3 is a histogram of every pixel color using the CIE (x, y) coordinate in 13,000 picture images stored in the Eastman Kodak library [7, 8]. It shows that most real-world colors are located in the center of the CIE coordinate, namely, near the white point. Figure 7.4 is another histogram that depicts the analysis of television image using the same method [9]. These results reveal the very interesting fact that most images can be expressed monochromatically (in white + black + gray) with some color addition [10].

The four-pixel approach (RGBW: four subpixels of red, green, blue, and white) utilizes this (monochrome) facility. (To avoid confusion with the RGB pixelation method, the RGB three-pixel method with white emission + color filter is denoted as W-RGB, and RGB four-pixel method with white emission + color filter is denoted as W-RGBW.)

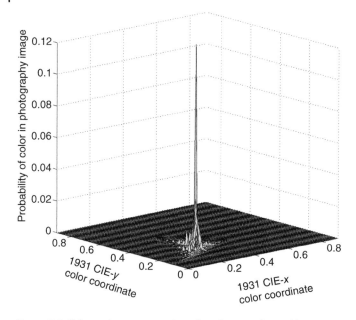

Figure 7.3 Schematic representation of a color-coordinated histogram constructed from analyses of 13,000 photographic images.

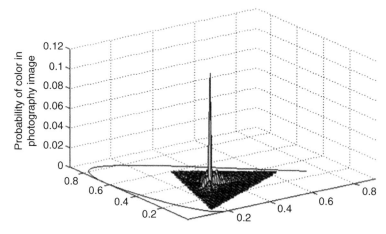

Figure 7.4 Schematic representation of a color-coordinated histogram constructed from analysis of a television signal.

With the W-RGBW method, white emission passes through a color filter with absorption for the R (red), G (green), and B (blue) subpixels. However, the white (W) subpixel does not cause absorption, so it can enable high efficiency and long lifetime. As discussed earlier, most real-world colors are near white, so the display power of a panel using the W-RGBW method does not change significantly even though its reproducible color gamut is wide. This ingenious strategy regarding knowledge and use of the W-RGBW method is an important breakthrough toward improving the power consumption of the white + color filter method and rendering this method competitive with the RGB pixelation method.

7.3.2 RGBW Driving Algorithm

With the RGBW driving method proposed by Murdoch et al., white emission features low-power driving capability without sacrificing color reproduction [11].

The conventional method uses red, green, and blue to display arbitrary colors (Figure 7.5). Murdoch's RGBW driving method also uses three primary colors to display arbitrary colors but selected the three primary colors according to the following rules:

- Color inside the B–G–W triangle is expressed by the three primary colors blue, green, and white, excluding red (Figure 7.6).
- Color inside the G–R–W triangle is expressed by the three primary colors green, red, and white, excluding blue (Figure 7.7).
- Color inside the R–B–W triangle is expressed by the three primary colors red, blue, and white, excluding green (Figure 7.8).

According to this approach, it is possible to fully utilize the advantage of high-efficiency white as well as the ability to display pure colors. Figure 7.9

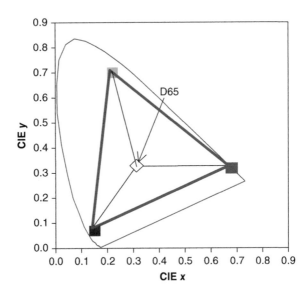

Figure 7.5 Graph illustrating color reproduction using the normal RGB method.

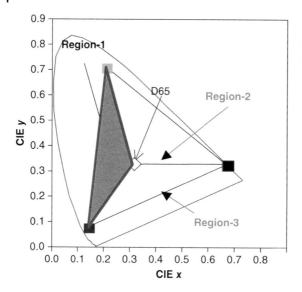

Figure 7.6 Plot showing color expression by GBW (green + blue + white) subpixels.

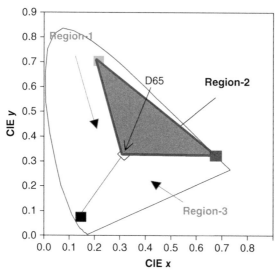

Figure 7.7 Plot showing color expression by RGW (red + green + white) subpixels.

shows the actual power consumption measurement results of W-RGB displays and W-RGBW displays using various images [8, 12]. The result shows that the W-RGBW four-subpixel method has almost half the power consumption of the W-RGB display panel. Keep in mind that that this discussion is *not* fully applicable to LCDs. The difference lies in the fact that OLED power consumption is proportional to the electric current, not aperture ratio, while an LCD's power consumption is proportional to the aperture ratio. In the LCD case, RGBW

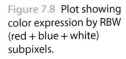

Figure 7.8 Plot showing color expression by RBW (red + blue + white) subpixels.

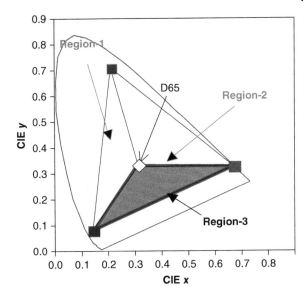

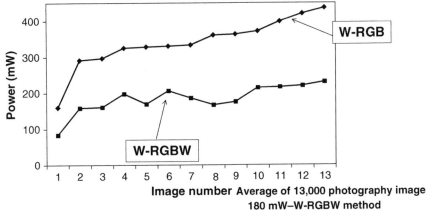

Figure 7.9 Graph comparing power consumption levels of W-RGB and W-RGBW methods for various images.

driving causes display transmittance loss due to narrower subpixel size by the use of four subpixels. Power reduction by RGBW driving is diminished by low transmittance of LCD pixels. In the case of OLED, luminous flux is almost the same regardless of aperture ratio if the electric current, namely, display luminance, is the same. So the merit of power consumption elimination can be brought about for OLED. The problem for OLED displays is the lifetime reduction due to aperture decrease.

Figure 7.10 14-in. active-matrix OLED display by white + color filter method.

Figure 7.10 shows the 14-in. prototype by white + color filter method. The display has 1280×768 resolution, RGBW four subpixel arrangement, 78% NTSC color gamut with peak luminance 500 cd/m^2 [9].

7.4 AN APPROACH TO ACHIEVE 100% NTSC COLOR GAMUT WITH LOW POWER CONSUMPTION USING WHITE + COLOR FILTER

7.4.1 Consideration of Performance Difference between W-RGB and W-RGBW Method

7.4.1.1 Issues of White + Color Filter Method for Large Displays

Though white + color filter method can significantly reduce power consumption by using W-RGBW four-pixel approach, compared with the case of W-RGB three-pixel approach, RGB-pixelation method has been already achieving more than 100% NTSC color gamut for cellular phone displays. Then, it was investigated whether there is any way to further relax the trade-off of white+color filter approach, so that both low power consumption and high color gamut such as 100% NTSC can be satisfied [4, 5].

7.4.1.2 Analysis of W-RGBW Approach to Circumvent Its Trade-off Situation

Though power consumption can be reduced with W-RGBW method as Murdoch pointed out, the display power consumption depends on the image content, so it was not easy to calculate it without actual experiment.

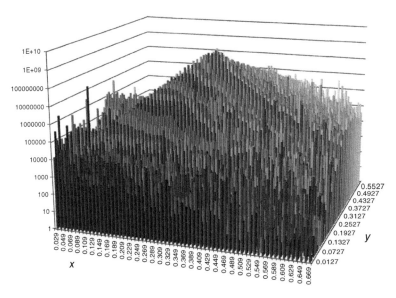

Figure 7.11 Histogram of pixel colors in 19,419 images (log scale).

Tsujimura [5] made a simulation on the 19,419 images and showed that the pixel color histogram (Figure 7.11) can be written as

$$(x - 0.31271)^2 + (y - 0.32902)^2 = [\{20.950 - \log_e(z)\}/40.516]^2 \qquad (7.3)$$

Then, the probability of having color coordinate (x, y) can be calculated as

$$P(xy) = z/N_{total} = \frac{1}{N_{total}} \exp(\log_e(z_{max}) - a\sqrt{(x - x_w)^2 + (y - y_w)^2}) \qquad (7.4)$$

where $N_{total} = 52{,}464{,}357{,}936$, $a = 40.516$, $\log_e(z_{max}) = 20.950$, $x_w = 0.31271$, and $y_w = 0.32902$.

Figure 7.12 shows the current density of (x, y) color coordinate of W-RGBW display using this statistics. Current density shows minimum value at white point and gradually increases as the color is saturated (larger chroma value). Especially, the current density shows very high value near blue color coordinate, as the current efficiency of blue is not very good, compared with other colors.

Though pure colors, especially blue, have large current consumption, the probability of having saturated colors is extremely low according to Figure 7.11. (Beware that the z axis of Figure 7.11 is logarithm scale.) To estimate the actual current consumption contribution in a display, Figure 7.11 needs to be multiplied by the value in Figure 7.12.

$$J_{PICTURE}(xy) = J_{TOTAL}(xy) \cdot P(xy) \qquad (7.5)$$

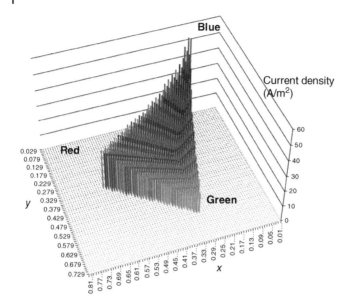

Figure 7.12 Total (W + R + G + B) driving current for W-RGBW method.

Equation 7.5 gives the histogram of current density for color coordinate (x, y) as shown in Figure 7.13. Pure colors give minor contributions and most of the current consumption is still made by near-white region.

When current efficiencies of 36.5 (cd/A) for D65 white, 7.6 (cd/A) red, 19.3 (cd/A) green, and 3.2 (cd/A) blue are assumed, by adding the current density for all the x and y, the total current density, 9.386 A/m², was calculated using probability distribution in Figure 7.11 for W-RGBW OLED display. Similarly, W-RGB OLED display (without white subpixel) was simulated and the current density was 21.060 A/m². According to this simulation, it was concluded that W-RGBW OLED display has almost half (1/2.2437) the current consumption of W-RGB OLED display. This corresponds to the observation of half power consumption by W-RGBW, discussed in Section 7.3.2.

As to the region distant from the white point, there is almost no contribution according to Figure 7.11 due to extremely low possibility of occurrence. In practice, the driving current does not change very much when the RGB subpixel efficiency is changed, as shown in Figure 7.14.

As R, G, B subpixel emissions do not affect the power consumption very much, if we can increase the white intensity without affecting the R, G, B emission intensity, the display power efficacy would be increased with no color-reproducibility handicap. However, R, G, B emission is made from white emission, so R, G, B emission should be proportional to the white emission if

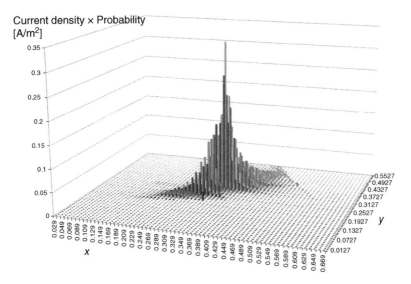

Figure 7.13 Current density of color coordinate (*x*, *y*) for actual display showing picture image.

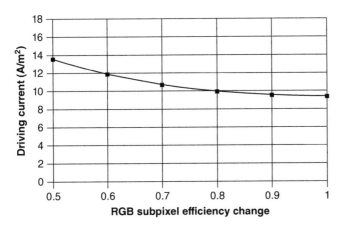

Figure 7.14 Driving current dependence of RGB subpixel efficiency in W-RGBW system.

white spectrum shape is kept the same. To overcome this situation, a trick can be applied.

Figure 7.15 shows the relationship between emission spectrum and color filter spectrum in the case of conventional design. To maximize the light output of OLED emission through color filter, the color filter peaks are normally matched with emission spectrum peak. A wide color filter pass band is used for each color so that a high transmittance and therefore brightness can be obtained for

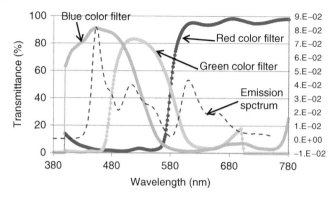

Figure 7.15 Conventional white + color filter approach.

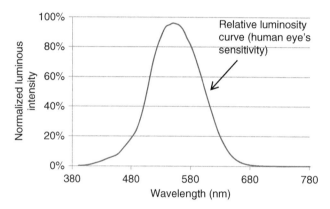

Figure 7.16 Relative luminosity curve.

each. However, the wide color filter spectrum results in desaturation of the colors and hence a poor color gamut, hence a trade-off exists between luminance and color gamut happens.

It is known that human eye's sensitivity has a peak at 555 nm wavelength as shown in Figure 7.16. If white OLED emission near the sensitivity peak is increased, the power consumption of W-RGBW display would be reduced, because most of the display power consumption is decided by white emission, as discussed. However, as 555 nm is between yellow and green, if such color appears in red, green, blue subpixels, color gamut would be reduced. Also for the high color gamut, narrow spectrum emission must be made. Therefore, it is necessary to adopt color filters with narrow spectrum width, which gives minimum overlap.

According to this analysis, two conclusions can be made for W-RGBW displays.

Conclusion 1: Current density is not very much affected by RGB subpixel efficiency. (Display power follows the same trend.)

Conclusion 2: By increasing the white subpixel efficiency, the display power would be further improved.

If both can be achieved, low power consumption can be combined with high color gamut capability on W-RGBW displays. This concept was checked through actual display fabrication.

7.4.1.3 Design of a Prototype to Demonstrate That Low Power Consumption Can Be Achieved with Large Color Gamut

To achieve low power consumption with high color gamut, an optimized set of OLED and color filter spectra was developed and applied to a W-RGBW AMOLED display [4].

R + G + B + Yellow OLED formulation used for the prototype is shown in Figure 7.17. The additional yellow spectrum component very well matches with relative luminosity curve, so the formulation makes much higher efficiency possible than conventional RGB-peak white.

The largest hurdle was color filter (Figures 7.18 and 7.19). Existing color filter for LCD could not show large color gamut for R + G + B + Yellow formulation, because the green and red color filters transmit light in the yellow wavelength region. Also the blue color filter spectrum was not narrow enough to achieve CIE-y = 0.06, which is necessary for Adobe-RGB/s-RGB color space reproduction. New color pigment and special milling/dispersion technology was applied so that the OLED panel demonstrated in IDW2008 [4] achieved NTSC 100% (Figure 7.20) with lower power consumption (less than 2W for 8-in., which corresponds to 64W for 32-in. television) than LCD display products on the market.

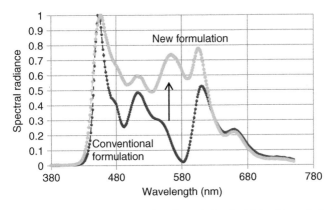

Figure 7.17 Wide spectrum white formulation for high white subpixel efficiency.

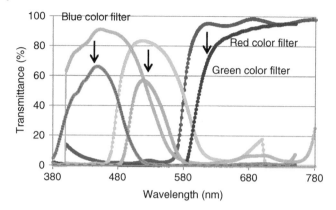

Figure 7.18 High color gamut color filter design for wide spectrum white emission.

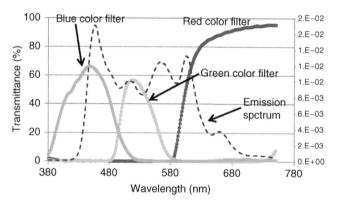

Figure 7.19 New formulation with new color filter.

The technology has been applied to state-of-the-art AMOLED displays on the market and is the dominant technology currently for the large AMOLED TVs.

7.4.1.4 Product-Level Performance Demonstration by the Combination of Scalable Technologies

Figure 7.21 a display prototype that has competitive performance as a display product including 100% NTSC color gamut, which was believed to be not achievable using the white + color filter method in combination with scalable technologies [4]. As blue is the most sensitive primary color (Figure 4.14 shows that the McAdam ellipse [the human eye perception limit] for blue is much smaller than for green or red), the blue pixel of the prototype follows $(x, y) = (0.15, 0.06)$, required for Adobe-RGB, s-RGB and Rec-709.

This 8.1-in. display has 1.5 W power consumption (300 cd/m^2 operation) using the W-RGBW approach. As 8-in. LCD displays have ~2–4 W power

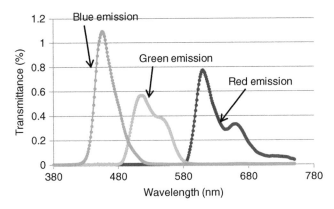

Figure 7.20 Emission output of W-RGBW display using new formulation, color filter set.

Figure 7.21 An OLED display prototype with a product-level specification using white + color filter technology (8.1-in. demonstration at SID 2008 by Eastman Kodak; this model has a 100% NTSC color gamut) [4].

consumption (data from specification sheet of eight commercially available 7~8-in. TFT LCD power consumption as of 2008), this white OLED prototype has already demonstrated competitive performance. For example, its power consumption corresponds to 64 W for a 32-in. television, and this is a sufficiently low value for a television set, which is typically 150 W in the case of an LCD television.

Table 7.3 Scalable TFT Technologies

Technology	Example	Advantage	Disadvantage
SLS-LTPS	Samsung	High mobility	ELA-related issues
SPC(MILC)	CMEL, Samsung, Sharp	Proven mass production	Leakage current and instability due to residual Ni; 6 masks for top gate
SPC(ASPC)	LG Display, Samsung	Low leak current due to, absence of residual Ni	6 masks for top gate; glass shrinkage
μc-Si by CVD	Samsung	4 masks	Low mobility, low deposition rate, large OFF current, necessity of compensation circuit
μc-Si by diode laser	Sony	5 masks, (etch-stopper type)	Large OFF current, low throughput, necessity of compensation circuit
Oxide TFT	Samsung, LG Display, Sony, Canon	4 or 5 masks, high mobility, uniformity, low photo leak current, high aperture ratio	Negative bias stress instability under illumination, necessity of compensation circuit

This prototype proved that the display specification according to the market requirement can be achieved by the combination of high-yield (>90%) technologies, therefore means that the OLED television industry can be realized with low risk of investment.

For this proof-of-concept prototype, scalable compensation technology and scalable TFT technology was used. Regarding the compensation circuit, the GMC method, discussed in Section 6.4.4, was used. As to the TFT, Table 7.3 shows a comparison of the risk profiles for each technology when the size becomes larger. Among all these technologies, SPC polysilicon technology was used, which has been already used in manufacturing. As to the detail of scalable TFT backplanes, the SLS method, microcrystalline and amorphous silicon method, solid-phase crystallization, and oxide semiconductors are discussed in Sections 6.5.1.1, 6.5.3–6.5.5, respectively.

References

1 E. Kelly, Considering color performance in curved OLED TVs, *Information Display Magazine*, Vol. 6, Issue 13, pp. 6-11 (2013).

2 M. Miller, M. Murdoch, J. Ludwicki, and A. Arnold, Determining power consumption for emissive displays, *SID Symposium Digest of Technical Papers*, Vol. 37, Issue 1, pp. 482–485 (2006).

3 R. Troutman, Forecasting array yields for large-area TFT LCDs, *SID 1990 Digest*, p. 197 (1990).

4 T. Tsujimura, S. Mizukoshi, N. Mori, K. Miwa, Y. Maekawa, M. Kohno, K. Onomura, K. Mameno, T. Anjiki, A. Kawakami, and S. Van Slyke, Scalable AMOLED technologies for TV application, *IDW 2008 Digest*, p. 145 (2008).

5 T. Tsujimura, High performance organic-light-emitting-diode television system for high yield manufacturing (Doctorial dissertation) (2015).

6 Y. Matsueda, *Low Cost OLED Device, Tech-On Cutting-Edge Display Technologies*, Nikkei Business Publications, 2009/11/24 (**2009**).

7 M. Miller, M. Murdoch, J. Ludwickia, and A. Arnold, Determining power consumption for emissive displays, *SID 2006 Digest*, p. 73 (2006).

8 A. D. Arnold, T. K. Hatwar, M. V. Hettel, P. J. Kane, M. E. Miller, M. J. Murdoch, J. P. Spindler, S. A. VanSlyke, K. Mameno, R. Nishikawa, T. Omura, and S. Matsumoto, Full-color AMOLED with RGBW pixel pattern, *Asia Display*, p. 809 (2004).

9 J. W. Hamer, A. D. Arnold, M. L. Boroson, M. Itoh, T. K. Hatwar, M. J. Helber, K. Miwa, C. I. Levey, M. E. Long, J. E. Ludwicki, D. C. Scheirer, J. P. Spindler, and S. A. Van Slyke, System design for a high color-gamut TV-sized AMOLED display, *J. SID* **16**:3 (2008).

10 World is grayish, *Nikkei Sangyo Newspaper*, 5/30/2008, p. 1 (2008).

11 M. J. Murdoch, M. E. Miller, and P. J. Kane, Perfecting the color reproduction of RGBW OLED, *Proceedings of the 30th Intnatl Congress of Imaging Science* (ICIS 2006) (2006).

12 J. P. Spindler, T. K. Hatwar, M. E. Miller, A. D. Arnold, M. J. Murdoch, P. J. Kane, J. E. Ludwicki, and S. A. Van Slyke, Lifetime- and power-enhanced RGBW displays based on white OLEDs, *SID 2005 Digest*, p. 36 (2005).

8

New OLED Applications

OLED technology has been used for existing applications such as cellular phones, MP3 players, and televisions, but OLED technology also has the potential to penetrate into new applications because

1. an all-solid device (when blanket encapsulation is used) is suitable for flexible applications;
2. use of a high-transmittance electrode enables transparent displays;
3. a thin display module is suitable for overlapped tiling.

8.1 FLEXIBLE DISPLAY/WEARABLE DISPLAYS

8.1.1 Flexible Display Applications

There are many flexible display [1–4] technologies reported in the OLED field (Figure 8.1). With flexibility properties, smart phone with round boarder (Figure 4.28), rollable (Figures 8.2 and 8.3) or foldable (Figure 8.4) displays can be fabricated. Also their robust, unbreakable properties may become important for future mobile applications.

8.1.2 Flexible Display Substrates

To make a flexible OLED device, there are three substrate technologies as shown in Table 8.1 [5, 6]. Glass substrate has perfect barrier property against oxygen and humidity. Recently, very thin glass (e.g., 0.1 mm thickness) has been reported. Such thin glasses can be handled by mounting the thin glass on thick carrier glass [7]. However, glass substrate is fragile and can be broken by the damages. Breakage during manufacturing would cause suspension of processing for hours and may bring about economical loss. Also breakage of the product may occur after purchase by the end user.

In the case of polymer substrate, thinness, light weight, and flexibility can be fully enjoyed. However, humidity and oxygen might cause reliability issues, such as dark spots or pixel shrinkage.

OLED Display Fundamentals and Applications, Second Edition. Takatoshi Tsujimura.
© 2017 John Wiley & Sons, Inc. Published 2017 by John Wiley & Sons, Inc.

Figure 8.1 An example of a flexible display (demonstration at FPD International 2007 by Samsung SDI).

Regarding the metal foil substrate, barrier property is as perfect as glass substrate. However, the metal foil normally has surface spikes, which may cause interlayer shorts between OLED electrodes. (Planarization layer reduces the issue.) Also metal foil is hard to handle on a roll-to-roll processing line, due to breakage and the weight of the rolls. Among these options, a plastic substrate is viable, if its barrier properties can be sufficiently improved. Such enhanced plastic substrates with barrier properties are referred to as barrier films.

Barrier films can be made by either roll-to-roll coating of barrier layers onto a plastic substrate or can be made by a laser liftoff process during device fabrication.

8.1.3 Laser Liftoff Process

Figure 8.5 shows the laser liftoff process presented by Miyasaka et al in 2007 [8]. LTPS-TFT is fabricated from an amorphous silicon layer, deposited on a glass substrate (Figure 8.5a). The device side of the substrate is glued by

Spindle   Rolled display   Flexible cable

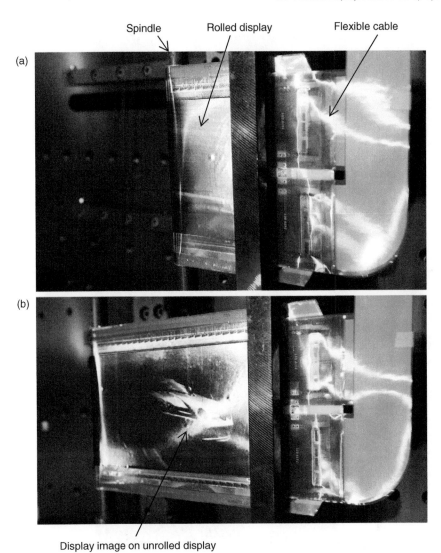

Display image on unrolled display

Figure 8.2 Demonstration of rollability of a flexible display: rolled (upper) and unrolled (lower) display (demonstration at SID2010 by Sony).

temporary adhesive to a temporary substrate (Figure 8.5b). XeCl excimer laser irradiation from the original glass substrate side breaks the adhesion of the TFT matrix to the original substrate and the devices are released (Figure 8.5c). The released device is glued to its final plastic substrate (Figure 8.5d). And finally, the temporary substrate is removed (Figure 8.5e).

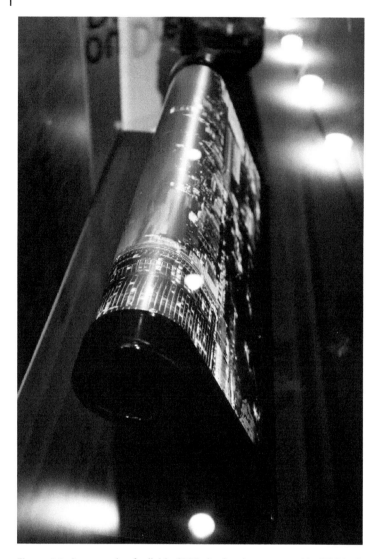

Figure 8.3 An example of rollable OLED display demonstrated in SID Display Week 2015 by LG display.

Since this laser transfer method was presented in 2001 [9], few such products have been reported; however, due to the improvement in barrier film technology to be applied to OLED display (discussed in Section 8.1.4), laser liftoff technology has suddenly come back under the spotlight. Curved cell phone displays fabricated by liftoff are gaining a lot of attention as product differentiators [10].

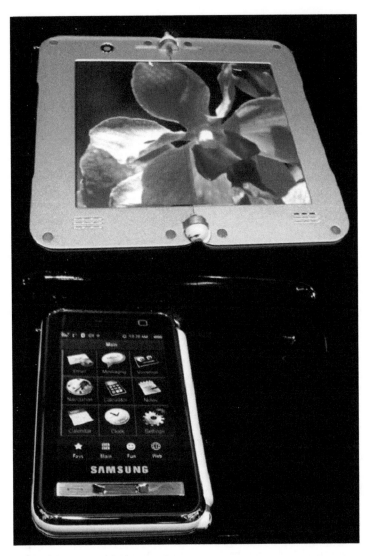

Figure 8.4 An example of a foldable display (demonstration at SID2008 by Samsung Mobile Display).

Although laser liftoff is an attractive method, which can enable flexible displays, cost is still high due to the complicated process. Many modifications have been proposed, including the use of solid-state lasers and other kinds of release layers.

Table 8.1 Various Substrate for Flexible OLED Device

	Pros	Cons
Thin glass	Barrier property Thinness Thermal durability	Breakage during manufacturing Product breakage
Plastic with barrier film	Flexibility Thinness Lightweight	Barrier property Thermal durability

Metal foil	Barrier property
	Thermal
	    conductivity
	Thinness
	Thermal durability
		Flatness (Needs insulator)
		Surface breakage in R2R
		    process
		Needs top emission

8.1.4 Barrier Technology for Flexible Displays

Plastic substrates are normally used for flexible displays, which require barrier capability integrated in the substrates. OLEDs are subject to damage by moisture and oxygen, so the substrate needs to have low permeability. Burrows et al. report that a water vapor transmission rate (WVTR) less than 10^{-6} g/m^2·day is necessary for OLED applications [11]. WVTR can be measured by a calcium test or by a vacuum method, as discussed in Section 3.3.5.4. A barrier layer is formed on the substrate using a similar method to the thin film encapsulation of an OLED device, such as by chemical vapor deposition of Si3N$_4$ (discussed in Section 3.3.2), or by deposition of multiple organic/inorganic layers (discussed in Section 3.3.5.1). Al$_2$O$_3$ deposition by ALD (Figure 3.36) is also used for barrier-layer formation. To obtain transparency of the barrier film for high light transmission, SiO$_x$N$_y$ is also used [12]. A plastic substrate has a low processing temperature limit, so it is important to keep the substrate sufficiently cool during the barrier-layer formation process. It should be noted that the maximum processing temperature must be much lower than the glass transition temperature (T_g) of the substrate. For example, the T_gs of the most popular substrates, such as polyethylene terephthalate (PET) and polyethylene naphthalate (PEN), are 150 and 180–220 °C, respectively, but the upper limit of the processing temperature needs to be lower, such as 80 °C for unannealed PET and 120 °C for unannealed PEN substrates [13]. (Annealing can increase the temperature limit to some extent, for example 120 °C for PET [14].)

There are also some examples of the use of metal foil substrates. A metal foil substrate [15] has merits in having high barrier capability and heat dissipation, which can extend the lifetime of an OLED device by achieving low-temperature operation due to the high thermal conductance of the metal electrode. A top-emission OLED structure must be used for a metal substrate.

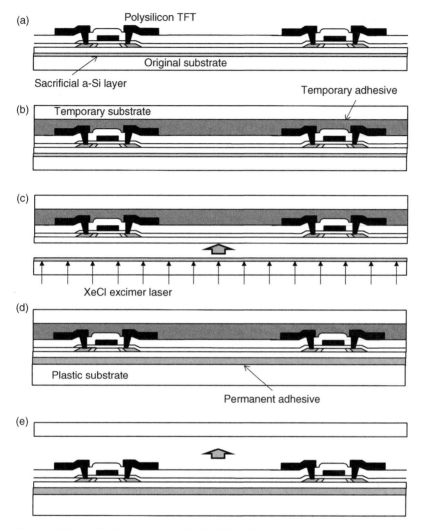

(a) Polysilicon TFT

Original substrate

Sacrificial a-Si layer

Temporary adhesive

(b) Temporary substrate

(c)

XeCl excimer laser

(d)

Plastic substrate

Permanent adhesive

(e)

Figure 8.5 Laser liftoff process to make flexible device.

Normally, a metal foil substrate has a rough surface, so a planarization layer must be deposited before fabricating the OLED device.

8.1.5 Organic TFTs for Flexible Displays

In the case of OLED fabrication on plastic substrates, a maximum processing temperature must not be exceeded, as already discussed. Normally, the maximum processing temperature of amorphous silicon TFTs is about 300–400 °C and that of LTPS TFT is about 500 °C. There is no plastic substrate that can survive such high-temperature processing. Too low a process temperature during silicon TFT formation degrades the quality of the amorphous silicon film, which causes many problems, such as high leakage current, low mobility, hysteresis, and instabilities, so it is necessary for flexible display processing to use a different approach from processing on glass substrates.

There are two popular approaches to circumvent these temperature issues, that is, organic TFTs or oxide TFTs. Oxide TFTs have already been discussed in Section 6.5.5, so only organic TFTs are discussed in this section.

8.1.5.1 Organic Semiconductor Materials

Figures 8.6 and 8.7 show examples of basic polymer conductive molecules and basic small-molecule conductive molecules, respectively. Three approaches to design a conductive organic molecule [16] are as follows:

1. Implement a long series of conjugated π bonds.
2. Stack planar molecules with π electrons.
3. Utilize the charge transfer interaction. (Figure 8.8 shows an example of a charge-transfer complex. See Section 2.6.2.2.)

The left-hand molecular structure in Figure 8.7 shows the most popularly used p-type organic semiconductor, pentacene. It is composed of five benzene rings and has a relatively long conjugated structure, which enables charge conduction. The right-hand structure in Figure 8.7 shows copper phthalocyanine, which is also known as a p-type organic semiconductor. (It is also used as hole injection material as discussed in Section 3.1.1.1.) Copper acts as an electron acceptor, so the phthalocyanine structure conducts by hole transport. Phthalocyanine can also conduct electrons by electron-acceptor modification of the phthalocyanine side chains [17].

Polyacetylene

Poly(phenylenevinylene) PPV

Polyaniline

Polythiophene

Figure 8.6 Examples of organic polymer conductors.

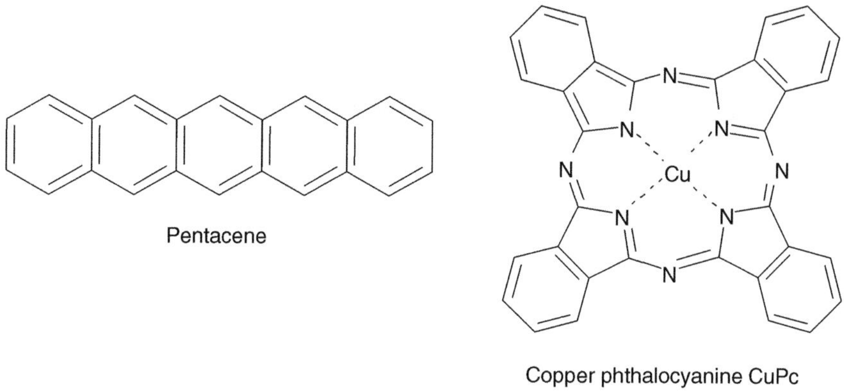

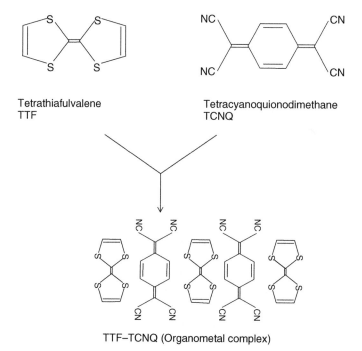

Pentacene

Copper phthalocyanine CuPc

Figure 8.7 Examples of small-molecule organic semiconductors.

Tetrathiafulvalene
TTF

Tetracyanoquionodimethane
TCNQ

TTF–TCNQ (Organometal complex)

Figure 8.8 An example of a charge transfer complex.

8.1.5.2 Organic TFT Device Structure and Processing

As discussed in Section 6.1, there are two types of TFTs: bottom gate and top gate. For bottom-gate organic TFTs, an additional variation is possible. Figure 8.9a shows the top-contact bottom-gate TFT, in which the lower

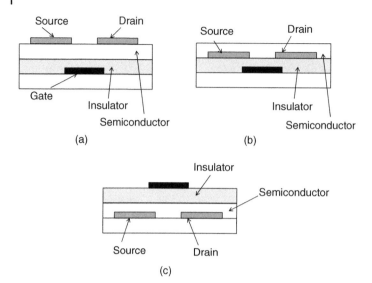

Figure 8.9 Various organic TFT structures. (a) Top-contact bottom gate TFT.
(b) Bottom-contact bottom gate TFT. (c) Top gate TFT.

surfaces of the source and drain electrodes contact the semiconductor layer; Figure 8.9b shows bottom-contact bottom-gate TFT, in which the upper surfaces of the source and drain electrodes contact the semiconductor layer. The top-contact type can be fabricated with minimum exposure of the semiconductor–insulator interface to the ambient environment, so a relatively high-quality channel layer can be formed, making this the most popularly used structure in organic TFTs. On the other hand, the bottom-contact type has the merit that the semiconductor layer does not go through many processing steps after it is formed, so processing damage can be minimized.

The channel layer of an organic TFT can be described as "a π electron conduction system formed by organic molecules coupled by weak interaction caused by Van der Waals forces" [18]. To achieve high conductivity in an organic semiconductor, it is very important to maximize the intermolecular orbital overlap of π electrons. Therefore, a highly crystallized film produces higher mobility. In the case of a polycrystalline film, grain mobility dominates conduction because the mobility at the grain boundary is lower than that within a grain. That is why grain size should be maximized by process optimization so that the fraction of the semiconductor layer occupied by grain boundaries is minimized.

Organic semiconductors are susceptible to oxidation, which causes instability. To avoid this problem, molecules with lower HOMO values, which reduce oxidation, have been developed.

8.1.5.3 Organic TFT Characteristics

The channel current of an organic TFT follows the gradual-channel approximation, the same principle discussed in Section 6.3. Current in the saturation region can be expressed as

$$I_D = \frac{\mu C_{OX}}{2} \frac{W}{L} (V_{GS} - V_{TH})^2 \tag{8.1}$$

where μ is mobility, V_{TH} is threshold voltage, C_{OX} is gate insulator capacitance per unit area, and W and L are channel width and channel length, respectively.

Normally, MOSFETs made with crystalline or polycrystalline TFTs keep a near-constant mobility value when the gate electric field is changed. However, in the case of organic TFTs, the mobility μ has a strong field dependency due to the hopping conduction mechanism, which was discussed in Section 2.3.7.3 (Poole–Frenkel conduction) and follows the equation as shown in (2.47).

$$\mu = \mu_1 \exp\left(-\frac{q}{kT}\beta\sqrt{E}\right) \tag{8.2}$$

where μ_1 is the mobility at zero electric field and β is the Poole–Frenkel factor.

8.2 TRANSPARENT DISPLAYS

Figure 8.10 shows an example of a transparent OLED display. Many companies are currently showing prototypes to explore possible markets for this new technology. They are targeting window information displays, head-mounted displays, digital speed meters, and so on. This new display technology may also open up novel applications.

To manufacture a transparent OLED display, a key technology is the transparent cathode electrode. With a conventional bottom emission structure, the cathode is made from a metal, such as aluminum. Metal is opaque, so it needs to be replaced with a transparent electrode. In the case of a top-emission structure, the cathode is made from a semitransparent metal, such as MgAg. By the combination of conventional technologies—an ITO anode, which is used for bottom emission, and a semitransparent metal cathode, which is used for top emission—a semitransparent OLED display can be fabricated. However, for an application such as a window information display, light transmittance and achromaticity are very important. A semitransparent metal cathode may not be sufficient to meet the market requirement.

To increase the transmittance, an ITO cathode is suitable. However, it is known that ITO sputtering can cause much damage to an OLED device. To circumvent this issue, the facing-target sputtering (FTS) system [19] provides a solution (Figure 8.11). FTS uses two sputtering targets face-to-face as a cathode and a magnetic field is applied perpendicular to the target faces (cf. the conventional DC sputtering system, discussed in Section 6.2.2, which

Figure 8.10 An example of a transparent OLED display (demonstration at FPD International 2009 by LG Display).

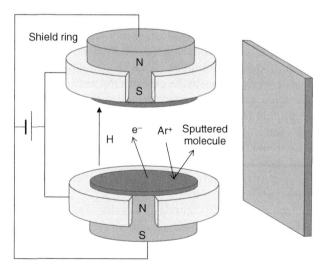

Figure 8.11 A schematic explanation of facing-target sputtering system.

has a configuration with the sputtering target and substrate face-to-face and a DC voltage is applied between target and substrate). Using magnets, positioned behind the target, damage-free deposition can be achieved because the magnetic and electric fields are both parallel to the target, so negative ions created by sputtering gas do not bombard the substrate. Using low-damage sputtering, high-quality transparent cathode electrodes can be fabricated [20].

8.3 TILED DISPLAY

Although widescreen displays are attractive for signage or multiple viewer applications, it is not possible to make displays larger than the limitations imposed by the mother glass or equipment size restrictions, such as the excimer laser radiation length. Also there are various needs for nonstandard displays with formats other than 4:3 or 16:9. Tiling is an attractive approach to realizing these requirements.

8.3.1 Passive-Matrix Tiling

When the display unit is tiled, discontinuity between the adjacent panels must be minimized. Therefore, the discontinuity must be below the level of the human eye's perception.

Three human visual perception limits are defined: (1) the minimum visible limit, (2) the minimum separable limit (hyperacuity), and (3) the minimum resolvable limit [21]. The angle subtended at the eye at the minimum resolvable limit distance from the observed object is called the minimum angle of resolution (MAR). Visual acuity is defined as the inverse of the minimum resolvable visual angle measured in minutes of arc. Therefore, viewing distance D, the minimum resolvable limit distance d_{MIN}, the minimum angle of resolution θ_{MAR}, and visual acuity V have the following relationship:

$$d_{MIN} = D \cdot \tan \left(\frac{2\pi}{V \times 21,600} \right) \tag{8.3}$$

Using this equation, the distance at which individual pixels can no longer be resolved can be calculated as given in Table 8.2 [22, 23].

For signage applications, viewing distance is much larger than that for television viewing at home, so pixels larger than 1 mm can be used with no problem, which makes the situation easier for tiling.

Figure 8.12 shows an actual implementation of tiled passive matrix display installed in a museum. Passive-matrix tiled panels to make a spherical OLED display [24]. Using a tiled arrangement of many small displays, the large screen achieved a curved profile using glass substrate. As this tiled display is intended for signage applications, it has a large 3 mm × 3 mm pixel pitch. (The panel

Table 8.2 Viewing Distance Required for Signage Display

Pixel Size (mm)	Viewing Distance Required (m)
1	3.4
2	6.9
3	10.3
4	13.8
5	17.2
6	20.6
7	24.1
8	27.5
9	30.9
10	34.4

unit size is 24 mm × 96 mm before tiling.) To minimize interpanel artifacts, the interpixel distance inside the panel is the same as the interpixel distance of adjacent panels ($a = b$), as shown in Figure 8.13.

Although visual artifacts can be reduced by minimizing the nonemission region between the panels, it is very difficult to make a tile boundary invisible, because the eye is extremely sensitive to luminance or spatial discontinuities at linear boundaries, so that a luminance change of 10% of the normal visual acuity limit is detectable [25]. Lowe et al. [25] report an LCD system that removes the nonemission boundary using a seamless image guide array (Figure 8.14). LCD pixel emission (with nonboundary region) is propagated through arrays of S-shaped image guides so that the emissive area covers the whole screen with the width of the nonemission regions being equal to the width of the gaps between individual image guides. Nevertheless, extreme care must be taken if the intertile boundaries are to be invisible. The method could be applied to OLED displays to enable a seamless tiling display.

8.3.2 Active-Matrix Tiling

Large display prototypes using active-matrix tiling also have been reported [26, 27] (Figure 8.15). The main purpose of tiling is to eliminate the shadow-mask and the excimer laser annealing restrictions.

As already discussed, there are other alternate technologies developed to circumvent these issues, so the necessity of tiling technology for active-matrix displays is decreasing. The remaining merit of active-matrix tiling would be the freedom to make custom display formats and very large displays.

Figure 8.12 An example of a tiled passive-matrix display (demonstration at FPD International 2011"Geo-cosmos" spherical display in Miraikan museum, Japan by Mitsubishi Electric).

Figure 8.16 shows an overlapped tiling method reported by Shim et al. [27]. In the reported structure, the use of an index-matched transparent plate and the reduction of the air gap between the transparent plate and substrate 1 play very important roles to avoid the loss of emission from substrate 2. Also, use of a thin display module is very important to avoid black lines caused by the gap between substrate 1 and substrate 2 when viewed from a positive angle.

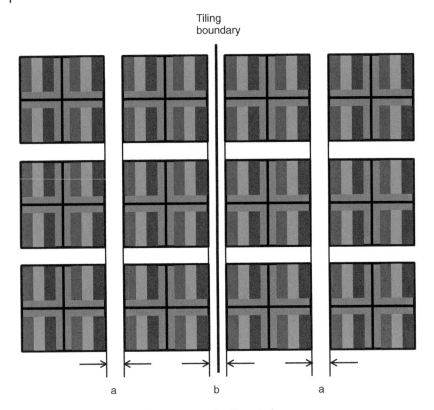

Figure 8.13 An example of a passive-matrix tiling pixel arrangement.

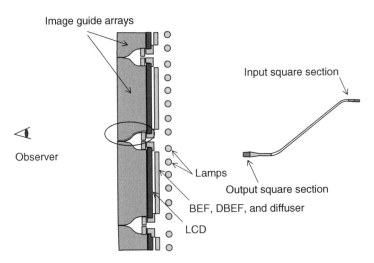

Figure 8.14 Seamless tiling using image guide arrays [25].

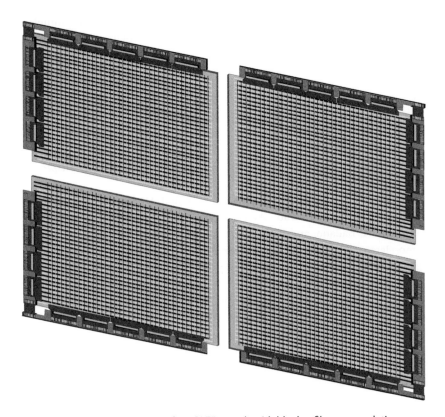

Figure 8.15 Tiling display using four OLED panels with blanket film encapsulation.

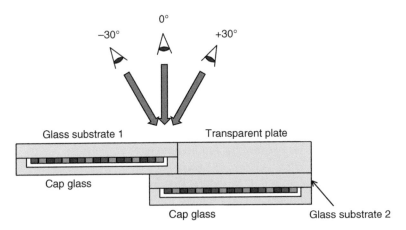

Figure 8.16 An example of active-matrix display tiling [27].

Using an overlapped tiling approach introduces an undesired asymmetric artifact, which obscures the pixel on substrate 2 from negative observation angles.

References

1 M. Hack, R.-Q. Ma, and J. J. Brown, Flexible OLEDs: challenges, opportunities, and current status, *IMID 2009 Digest*, p. 211 (2009).

2 A. Nishioka, OLED full-color panels for mobile application, *Asia Display 2004 Digest* (2004).

3 A. Krasnov, P. Hofstra, D. Johnson, and R. Wood, High-contrast black layer OLEDs on flexible substrate, *Asia Display/IDW'01 Digest*, p. 1415 (2001).

4 S.-J. Lee, A. Badano, and J. Kanicki, Monte Carlo simulations and opto-electronic properties of polymer light-emitting devices on flexible plastic substrates, *IDRC 2003 Digest*, p. 26 (2003).

5 T. Tsujimura, J. Fukawa, K. Endoh, Y. Suzuki, K. Hirabayashi, and T. Mori, Flexible OLED using plastic barrier film and its roll-to-roll manufacturing, *SID 2014 Digest*, Vol. **45**, Issue 1, pp. 104–107 (2014).

6 T. Tsujimura, J. Fukawa, K. Endoh, Y. Suzuki, K. Hirabayashi, and T. Mori, Development of flexible organic light-emitting diode on barrier film and roll-to-roll manufacturing, *J. SID* **22**(8):412–418 (2014).

7 T. Higuchi, Y. Matsuyama, K. Ebata, and D. Uchida, A novel handling method of ultra-thin glass for thin and flexible displays, *SID 2012 Digest*, Vol. **43**, Issue 1, pp. 1372–1374 (2012).

8 M. Miyasaka, Suftla flexible microelectronics on their way to business, *SID 2007 Digest*, Vol. **38**, Issue 1, pp. 1673–1676 (2007).

9 S. Utsunomiya, T. Saeki, S. Inoue, and T. Shimoda, Low temperature poly-si TFT-LCD transferred onto plastic substrate using surface free technology by laser ablation/annealing (SUFTLA®), *Asia Display/IDW 01 Tech. Digest*, p. 339 (2001).

10 S. K. Hong, C. Jeon, S. Song, J. Kim, J. Lee, D. Kim, S. Jeong, H. Nam, J. Lee, W. Yang, S. Park, Y. Tak, J. Ryu, C. Kim, B. Ahn, and S. Yeo, Development of commercial flexible AMOLEDs, Vol. **45**, Issue 1, pp. 334–337 (2014).

11 P. E. Burrows et al., Gas permeation and lifetime tests on polymer-based barrier coatings, *SPIE Proc.* **4105**:75–83 (2001).

12 A. Sugimoto, A. Yoshida, and T. Miyadera, Development of a film-type organic EL electroluminescent (OEL) display, *Pioneer R&D*, **11**(3):48 (2007).

13 H. Klauk, *Organic Electronics: Materials, Manufacturing, and Applications*, John Wiley & Sons, p. 166 (2007).

14 H. Zhang and I. M. Ward, Kinetics of hydrolytic degradation of poly(ethylene naphtalene-2,6-dicarboxylate), *Macromolecules* **28**:7622–7629 (1995).

15 S. Chung, J.-H. Lee, J. Jeong, J.-J. Kim, and Y. Hong, Substrate thermal conductivity effect on heat dissipation and lifetime improvement of organic light-emitting diodes, *Appl. Phys. Lett.* **94**:253302 (2009).

16 N. Ogata, *Organic Conductive Polymers*, Kodansha Scientific, p. 1 (1990).

17 T. Morita et al., *Research of Ambipolar Organic Field-Effect Transistors*, Kyushu Institute Technology Academic Repository, p. 20 (2010).

18 Japan Society for the Promotion of Science, *Organic Semiconductor Devices*, Ohmsha, p. 189 (2010).

19 M. Naoe, S. Yamanaka, and Y. Hoshi, Facing targets type of sputtering method for deposition of magnetic metal films at low temperature and high rate, *IEEE Trans. Magnet* **28**:646 (1980).

20 S. Dangtip, Y. Hoshi, Y. Kasahara, Y. Onai, T. Osotchan, Y. Sawada, and T. Uchida, Study of low power deposition of ITO for top emission OLED with facing target and RF sputtering systems, *J. Phys.: Conf. Ser.* **100**:042011 (2008).

21 G. von Noorden and E. C. Campos, *Binocular Vision and Ocular Motility: Theory and Management of Strabismus*, Mosby, p. 114 (2001).

22 M. Aston, Design of large-area OLED displays utilizing seamless tiled components, *J. SID* **15**(8):535 (2007).

23 M. Aston, Design of large area OLED displays utilising seamless tiled components, *IDRC 2006 Digest*, p. 211 (2006).

24 Z. Hara, K. Maeshima, N. Terazaki, S. Kiridoshi, T. Kurata, T. Okumura, Y. Suehiro, and T. Yuki, The high performance scalable display with passive OLEDs, *SID 2010 Digest*, p. 357 (2010).

25 A. C. Lowe, N. A. Gallen, and P. A. Bayley, Tiling technology for large-area direct-view displays, *J. SID* **14**(5):427 (2006).

26 S. Terada, G. Izumi, Y. Sato, M. Takahashi, M. Tada, K. Kawase, K. Shimotoku, H. Tamashiro, N. Ozawa, T. Shibasaki, C. Sato, T. Nakadaira, Y. Iwase, T. Sasaoka, and T. Urabe, A 24-inch AM-OLED display with XGA resolution by novel seamless tiling technology, *SID 2003 Digest*, p. 1463 (2003).

27 H. S. Shim, I. S. Kee, Y. GuLee, Y. W. Jin, and S. Y. Lee, Simulation study for seamless imaging of OLED tiled display, *IDW 2008*, p. 173 (2008).

9

OLED Lighting

Although the efficacy of OLED lighting is still behind that of LEDs, there are several merits of OLED for lighting. As OLED is planar lighting source, the light is much softer than LEDs, which has concentrated bright light from the point source. Also the diffused light of OLED can be very close to the task surface without creating glare to the users [1].

The white-emission OLED device used for displays is very similar to the OLED lighting device (Figures 9.1 and 9.2), so the display techniques are mostly applicable in lighting applications. However, the specification for a lighting application should differ from that for a display. In particular, the light source must provide adequate color reproduction of the irradiated object. This metric is called the color rendering index (CRI).

9.1 PERFORMANCE IMPROVEMENT OF OLED LIGHTING

Figure 9.3 shows efficacy of commercial LEDs in the market [1]. The efficacy is improving year by year. Also it can be seen that the cool package LEDs have higher efficacy than warm package. Figure 9.4 shows the efficacy trend of OLED. When compared with Figure 9.3, white OLED efficacy is improving at a faster pace than LEDs.

The efficacy of fluorescent lamps on the market is around 60 lm/W, so OLED efficacy is already competitive with such conventional technologies. There is another aspect to have the best quality of light. The CRI discussed in the following section is an important metric, which quantifies apparent changes in the color of objects illuminated by a source, compared to the same objects illuminated by natural broadband sources such as daylight.

OLED Display Fundamentals and Applications, Second Edition. Takatoshi Tsujimura.
© 2017 John Wiley & Sons, Inc. Published 2017 by John Wiley & Sons, Inc.

Figure 9.1 An example of table lighting using OLED (demonstration at FPD International 2009 by Rohm).

Figure 9.2 An example of an OLED ceiling lamp (demonstration at FPD International 2009 by Lumiotec).

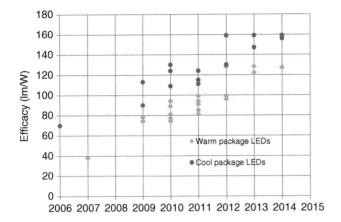

Figure 9.3 Efficacy of commercial LED packages.

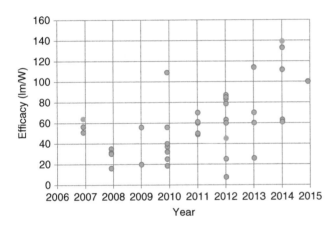

Figure 9.4 White OLED efficacy trend (extracted from SID digest database with keyword "lm/W OLED lighting").

9.2 COLOR RENDERING INDEX

The CRI is widely used to demonstrate the color reproduction of reflected light in lighting applications.

For the CRI calculation, the CIE1964 uniform color space (U^*, V^*, W^*) coordinate system is used (this color coordinate applies only to the CRI calculation):

$$W^* = 25Y^{1/3} - 17 \tag{9.1}$$

$$U^* = 13W^*(u - u_n) \tag{9.2}$$

$$V^* = 13W^*(v - v_n) \tag{9.3}$$

Here, the coordinate set (u, v) is

$$
\begin{aligned}
u &= \frac{4X}{X + 15Y + 3Z} \\
&= \frac{4x}{-2x + 12y + 3} \\
v &= \frac{6Y}{X + 15Y + 3Z} \\
&= \frac{6y}{-2x + 12y + 3}
\end{aligned}
\tag{9.4}
$$

where u_n and v_n are color coordinates of the light source.

As the color difference of the object between the standard light source and the sample light source are perceptually compensated according to color adaptation theory, so the following equations are applied:

$$
\begin{aligned}
u'_k &= u_r \\
v'_k &= v_r \\
u'_{ki} &= \frac{10.872 + 0.404\frac{c_r c_{ki}}{c_k} - 4\frac{d_r d_{ki}}{d_k}}{16.518 + 1.481\frac{c_r c_{ki}}{c_k} - \frac{d_r d_{ki}}{d_k}} \\
v'_{ki} &= \frac{5.520}{16.518 + 1.481\frac{c_r c_{ki}}{c_k} - \frac{d_r d_{ki}}{d_k}}
\end{aligned}
\tag{9.5}
$$

where (u_k, v_k) is the color coordinate of the sample light source and (u_{ki}, v_{ki}) is the color coordinate of each (i)th test coupon when illuminated by the sample light source.

Coefficients c and d can be expressed as follows:

$$
\begin{aligned}
c &= \frac{(4.0 - u - 10.0v)}{v} \\
d &= \frac{1.708v + 0.404 - 1.481u}{v}
\end{aligned}
\tag{9.6}
$$

Two color rendering values are commonly used: the general CRI (R_a) and the special CRI (R_i, where $i = 1{-}15$).

The spectrum distribution $S_r(\lambda)$ and $S_k(\lambda)$ for a standard light source and a sample light source is used to calculate the tristimulus value (X_r, Y_r, Z_r) and (X_k, Y_k, Z_k), respectively. For the ith sample color, tristimulus values for both light sources can be calculated as (X_{ri}, Y_{ri}, Z_{ri}) and (X_{ki}, Y_{ki}, Z_{ki}), respectively. Using

Eqs. (9.4)–(9.6), we obtain

$$W_{ri}^* = 25(Y_{ri})^{1/3} - 17$$
$$U_{ri}^* = 13W_{ri}^*(u_{ri} - u_r)$$
$$V_{ri}^* = 13W_{ri}^*(v_{ri} - v_r)$$
$$W_{ki}^* = 25(Y_{ki})^{1/3} - 17 \tag{9.7}$$
$$U_{ki}^* = 13W_{ki}^*(u_{ki} - u_k)$$
$$V_{ki}^* = 13W_{ki}^*(v_{ki} - v_k)$$

Therefore, color difference (variation) ΔE_i ($i = 1, 2, 3, \ldots, 15$) in the CIE1964 color coordinate system can be calculated as follows:

$$\Delta E_i = \sqrt{(U_{ri}^* - U_{ki}^*)^2 + (V_{ri}^* - V_{ki}^*)^2 + (W_{ri}^* - W_{ki}^2)^2}. \tag{9.8}$$

The special CRI R_i is

$$R_i = 100 - 4.6\Delta E_i \tag{9.9}$$

and the general CRI R_a can be expressed as

$$R_a = \frac{1}{8}\sum_{i=1}^{8} R_i \tag{9.10}$$

A large CRI gives vivid, natural colors. Normally, blackbody radiation spectrum is used for the standard light source. According to Eqs. (9.8)–(9.10), color rendering indices show the color difference between the color made by the sample source and the color made by the standard light source; therefore, a spectrum closer to the blackbody radiation needs to be achieved for a high CRI value. This is why white OLED or white emission by LED + fluorescence shows a higher CRI than a white light made by a peaked LED emission or a peaked laser emission. In the case of display, a primary color with a broader spectrum causes narrower color reproduction of a display (calculation for the display is discussed in Section 4.2.3), so attention should be paid to the difference between lighting and display.

9.3 OLED LIGHTING REQUIREMENT

As OLED lighting [2–4] is an emerging technology, customer requirements are not yet fixed and will change over time. As to the current general requirements, the ENERGY STAR® document, prepared by the US Department of Energy, provides useful information. Here we discuss the specifications outlined in the ENERGY STAR program *Requirements for Solid-State Lighting Luminaires Eligibility Criteria*, version 1.2 (Final Criteria).

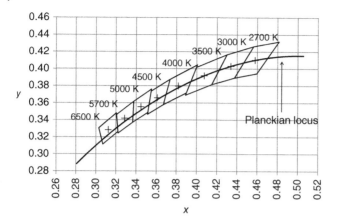

Figure 9.5 Graph showing eight nominal correlated color temperature (CCT) quadrangles on a CIE1931 chromaticity diagram.

9.3.1 Correlated Color Temperature (CCT)

As discussed in Section 4.2.5, the white color on a Planckian locus can be expressed by color temperature. However, the actual color coordinate of white emission may not be exactly on the Planckian locus. To express the color of near-white color emission not on the Planckian locus, the correlated color temperature (CCT) is used, defined as the color temperature of the (u, v) color coordinate on the Planckian locus, which is closest to the emissive color.

For CCT calculation, the CIE1976 (discussed in Section 4.2.3) color space is used, with which the color can be calculated using the following equations:

$$u' = \frac{4x}{-2x + 12y + 3} \tag{9.11}$$

$$v' = \frac{9y}{-2x + 12y + 3} \tag{9.12}$$

The distance from the Planckian locus on the CIE 1976 (u', v') coordinates is referred to as Δuv.

In the ENERGY STAR document, it is noted that a luminaire should have a CCT of 2700, 3000, 3500, 4000, 4500, 5000, 5700, or 6500 K.

The document also mentions that the luminaire must fall within the specified seven-step chromaticity quadrangles, defined by the seven-step (seven times standard deviation) McAdam ellipse (see Table 9.1 and Figure 9.5).

The human eye is very sensitive to any deviation from the blackbody line (described as the Planckian locus in Figure 9.5), so it is important to design a device formulation with a white point as close to the blackbody line as possible.

Table 9.1 CIE Coordinates of Nominal CCTs

	2700 K		3000 K		3500 K		4000 K		4500 K		5000 K		5700 K		6500 K	
	x	y	x	y	x	y	x	y	x	y	x	y	x	y	x	y
Center point	0.4578	0.4101	0.4338	0.4030	0.4073	0.3917	0.3818	0.3797	0.3611	0.3658	0.3447	0.3553	0.3287	0.3417	0.3123	0.3282
Tolerance quadrangle	0.4813	0.4319	0.4562	0.4260	0.4299	0.4165	0.4006	0.4044	0.3736	0.3874	0.3551	0.3760	0.3376	0.3616	0.3205	0.3481
	0.4562	0.4260	0.4299	0.4165	0.3996	0.4015	0.3736	0.3874	0.3548	0.3736	0.3376	0.3616	0.3207	0.3462	0.3028	0.3304
	0.4373	0.3893	0.4147	0.3814	0.3889	0.3690	0.3670	0.3578	0.3512	0.3465	0.3366	0.3369	0.3222	0.3243	0.3068	0.3113
	0.4593	0.3944	0.4373	0.3893	0.4147	0.3814	0.3898	0.3716	0.3670	0.3578	0.3515	0.3487	0.3366	0.3369	0.3221	0.3261

Table 9.2 lists the general requirements specified by the ENERGY STAR program. It should be noted that the lifetime of a panel is defined by 70% luminance: T_{70}.

Table 9.3 lists the near-term application requirements (category A). Here, luminaire efficacy is defined as follows:

$$\eta_{lum} = \frac{P_{lum,out}}{P_{lum,in}} \tag{9.13}$$

where

η_{lum} = luminaire efficacy

$P_{lum,out}$ = luminaire light output (including fixture efficacy and thermal effects)

$P_{lum,in}$ = luminaire input power

To satisfy the entire application requirement presented in Table 9.3, an efficacy of 35 lm/W is desired for near-term application. The ENERGY STAR document also mentions that 70 lm/W is the aim in the future (category B).

9.4 LIGHT EXTRACTION ENHANCEMENT OF OLED LIGHTING

Ecology has become a matter of global interest recently, as wasted energy causes multiple issues, such as global warming, depletion of oil resource and air pollution. Due to such reasons, high-efficiency lighting technology has been desired. OLED lighting is also subjected to such requirements. To improve the luminous efficacy, various light extraction methods are used.

9.4.1 Various Light Absorption Mechanisms

As discussed in Section 2.4.3, the major reasons for OLED to lose emission energy are surface plasmons, waveguided-mode, and substrate-mode absorption.

The following sections illustrate how to reduce light absorption related to such major issues.

There is also important thing to be considered for high efficiency. During the trial to reduce the emissive energy loss, light released from surface plasmon absorption can be trapped by waveguided or substrate mode. Also light released from waveguided mode can be trapped by substrate mode. It is quite important to consider how reduced loss mechanism leads to the actual output.

Table 9.2 General ENERGY STAR Program Requirements

All Luminaires	**Color spatial uniformity**	**Within 0.004 from weighted average point on CIE1976 (u', v') diagram**
	Color maintenance	**Within 0.007 on CIE1976 (u', v') diagram over lifetime**
	Color rendering index	**Indoor luminaire shall have a minimum CRI of 75**
Modules/arrays	L_{70}	Residential indoor: 25,000 h
		Residential outdoor: 35,000 h
		All commercial: 35,000 h
Power supplies	Power factor	Residential $\geq$ 0.70
		Commercial $\geq$ 0.90
	Minimum operating temperature	Minimum operating temperature $-20°C$ or below when used in luminaires intended for outdoor applications
	Maximum measured power supply case or manufacturer designated temperature Measurement point (TMPPS) Temperature	Not to exceed the power supply manufacturer maximum recommended case temperature or TMP when measured during in situ operation
	Output operating frequency	≥ 120 Hz
	Electromagnetic and radio frequency interference	Power supplies designated by the manufacturer for residential applications must meet FCC requirements for consumer use Power supplies designated by the manufacturer for commercial applications must meet FCC requirements for nonconsumer use
	Noise	Power supply shall have a Class A sound rating
	Transient protection	Power supply shall comply with IEEE C.62.41-1991, Class A operation. The line transient shall consist of seven strikes of a 100 kHz ring wave, 2.5 kV level, for both common mode and differential mode

Table 9.3 Near-Term ENERGY STAR Application Requirements

Definition	Luminaire efficacy	Luminaire light output (includes fixture efficiency and thermal effects)/Luminaire Input Power				

Category A: Near-term Applications

Residential Application

Definition	Under-cabinet kitchen lighting	Portable desk task lights	Recessed downlights and pendant-mounted downlights	Ceiling-mounted luminaires with diffusers	Cove lighting	Surface-mounted luminaires with directional head(s)
Minimum light output	125 lm (initial)	200 lm (initial)	≤ 4.5″ Aperture: 345 lumens (initial) > 4.5″ Aperture: 575 lumens (initial)	≤ 8″ Aperture: 375 lumens (initial) > 8″ Aperture: 750 lumens (initial)	200 lm (initial)	200 lm (initial)
Minimum luminaire efficacy	24 lm/W	29 lm/W	35 lm/W	30 lm/W	45 lm/W	35 lm/W
Allowable CCTs	2700, 3000, and 3500 K	2700, 3000, 3500, 4000, 4500, and 5000 K	2700, 3000, and 3500 K	2700, 3000, and 3500 K	2700, 3000, and 3500 K	2700, 3000, and 3500 K

Definition	Outdoor wall-mounted porch lights	Outdoor step lights	Outdoor pathway lights	Outdoor pole/arm-mounted decorative luminaires
Minimum light output	150 lm (initial)	50 lm (initial)	100 lm (initial)	300 lm (initial)
Minimum luminaire efficacy	24 lm/W	20 lm/W	25 lm/W	35 lm/W

	Outdoor wall-mounted porch lights	Outdoor step lights	Outdoor pathway lights	Outdoor pole/arm-mounted decorative luminaires
Minimum light output	150 lm (initial)	50 lm (initial)	100 lm (initial)	300 lm (initial)
Minimum luminaire efficacy	24 lm/W	20 lm/W	25 lm/W	35 lm/W
	Recessed downlights and pendant-mounted downlights	Under-cabinet shelf-mounted task lighting	Portable desk task lights	Wall wash luminaires
Minimum light output	≦ 4.5″ Aperture: 345 lumens (initial); > 4.5″ Aperture: 575 lumens (initial)	125 lm (initial)	200 lm (initial)	575 lm (initial)
Minimum luminaire efficacy	35 lm/W	29 lm/W	29 lm/W	40 lm/W
Allowable CCTs	2700, 3000, 3500, 4000, 4500, and 5000 K	2700, 3000, 3500, 4000, 4500, and 5000 K	2700, 3000, 3500, 4000, 4500, and 5000 K	2700, 3000, 3500, 4000, 4500, and 5000 K
	Bollards			
Minimum light output	Luminaire shall deliver < 15% of total lumens in the 90 – 110° zone and emit no light over 110°.			
Minimum luminaire efficacy	35 lm/W			

Category B: future performance targets

Luminaire efficacy	≧ 70 lm/W

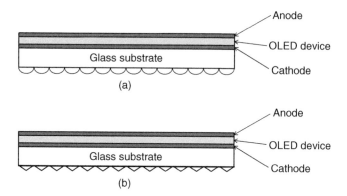

Figure 9.6 Schematic representations of (a) a microlens array and (b) a micropyramidal structure.

9.4.2 Microlens Array Structure

Figure 9.6 presents diagrams of the structures of a microlens array and a micropyramid used for light extraction.

To most effectively extract OLED emission light which propagates in a substrate, using a lens, the dimensions of the lens must be much greater than those of the emission area so that the lens covers all the luminous flux emitted from the area [5]. However, with this approach, the OLED emission area needs to be small enough compared with the lens area (e.g., to estimate the total amount of luminous flux from the emission coupon, a hemispheric lens with a much larger area than the emission coupon is used). In actual lighting panel design, it is not realistic to place a large lens structure over the emission area, so this approach is seldom used for an actual product. Instead, a microlens array with small diameters such as 10 μm is used. This light extraction method is using a different approach, such as a combination of light diffusion and refraction on the substrate surface.

9.4.3 Diffusion Structure

To extract the light in a substrate, a diffuser structure is also effective. Figure 9.7 shows the spin-coat-type diffusion layer that was reported by Tyan et al. [6].

Device A has two kinds of light scattering layers, external extraction structure (EES) at the glass–air boundary and internal extraction structure (IES) at the OLED device–glass boundary, whereas device B has EES only at the glass–air boundary. Among the total light flux emitted in the OLED device, there is (1) some that cannot exit the OLED device (waveguided mode, discussed in Section 2.4.3; waveguided-mode light is lost due to the energy loss during multiple reflection or light extraction at the OLED device edge),

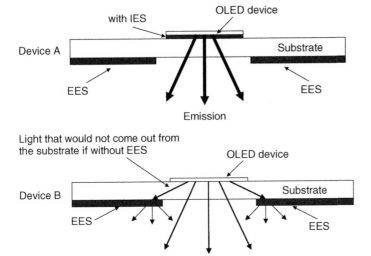

Figure 9.7 Diagrams of an internal extraction structure (IES) and an external extraction structure (EES).

(2) some that is lost due to plasmon absorption (light absorption caused by the coupling between plasmon, a type of electron vibration in metal, and the electromagnetic field of the light), (3) some successfully extracted from the OLED device, but which cannot exit the substrate (substrate mode, discussed in Section 2.4.3; substrate-mode light is also lost due to multiple reflection or extracted at the substrate edge), or (4) some extracted to the air and which reaches the observers (air mode). EES decreases the substrate-mode light and IES decreases the waveguided-mode light.

Figure 9.8 shows emission from an actual device. With IES, the intensity of the light emitted from the OLED device area is high due to light extraction from the waveguided mode. On the other hand, with only EES, emission intensity is lower in the device area due to the light loss caused by the waveguided mode, while light emission in the area surrounding the device occurs due to a larger amount of light propagated in the OLED device; some of the propagated light is extracted to the surrounding area by means of the EES structure.

Table 9.4 shows the actual experimental data. The device with EES has 1.92 times the external quantum efficiency of the non-EES device, and the device with IES has 2.28 times the external quantum efficiency. (This method is applicable for lighting but not suitable for displays because the diffusion capability of extraction layers causes image blurring and color mixture problems in the case of display applications.)

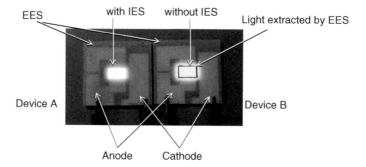

EES with IES without IES Light extracted by EES

Device A Device B

Anode Cathode

Figure 9.8 Image showing actual emission using IES and EES techniques.

Table 9.4 Light Extraction Experimental Data

Structure	Current Density (mA/cm²)	Luminance (cd/m²)	External Quantum Efficiency	Current efficiency (cd/A)	Color coordinate x	Color coordinate y	Voltage (V)	Power efficacies (lm/W)
—	5	651	5.7	13.0	0.493	0.434	3.1	13.0
EES only	5	1269	11.0	25.4	0.492	0.430	3.3	24.1
IES only	5	1430	13.0	28.6	0.475	0.408	3.1	29.0

9.4.4 Diffraction Structure

Diffraction structure is also used to extract the substrate-mode light. Figure 9.9 depicts the light extraction enhancement method using diffraction structure. After preparation of the diffraction structure and the high-refractive-index layer including compounds such as TiO_2 or SiN_x, the OLED device is fabricated. Periodical diffraction structure is fabricated between the substrate and the high-refractive-index layer. In the case of normal diffraction lattices in other industries, the pitch is designed so that the light path length difference between adjacent projections is equal to an integer times the wavelength of a monochromatic light. However, in the case of OLED lighting, white emission intensity with multiple wavelengths must be enhanced. To make this happen, a shorter lattice pitch close to or less than the visible-light wavelength is used. In such cases, light is scattered according to Mie scattering theory, which has minimal wavelength dependence.

9.4.5 Reduction of Plasmon Absorption

9.4.5.1 Plasmonic Loss Mechanism

Figure 9.10 shows the so-called Surface-plasmon polariton wave (SPP wave). It is a charge density oscillation, which propagates along the metal–dielectric interface.

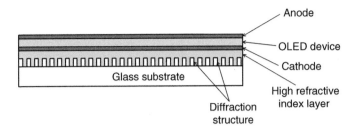

Figure 9.9 Schematic representation of light extraction enhancement employing diffraction structure.

Figure 9.10 Surface-plasmon polariton wave.

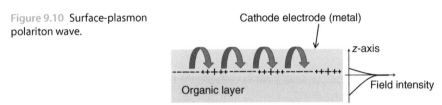

Electric dipoles associated with the OLED emitters couple with this SPP wave and lose a significant amount of energy. The electric field of the SPP is a so-called evanescent field, which decays rapidly as the distance from the dielectric–metal interface is increased.

Meerheim et al. made a quantitative study of energy loss in OLEDs by simulation [7] and showed that the surface plasmon loss can be significantly reduced by increasing the thickness of the ETL, which lies between the SPP wave and the radiative emitters. It was also concluded that the major portion of the loss is converted to a waveguided mode unless an appropriate outcoupling method is combined with surface plasmon loss reduction.

Although ETL thickness increase can reduce the plasmon loss, thick ETL causes larger voltage drop, which reduces the OLED device efficiency. Kato et al. combined ETL thickness increase with high mobility ETL and Internal Extraction Structure, and achieved very high white OLED efficacy 139 lm/W [8].

Other than the ETL thickness increase, plasmon loss can be reduced by horizontally oriented dipoles [9] (discussed in Section 2.4.3.4) or random structure in metal–dielectric contact [10, 11].

9.5 COLOR TUNABLE OLED LIGHTING

There are two types of color tunable OLED lighting devices as shown in Figure 9.11. The left structure has red, green, and blue lines, spatially allocated by mask patterning method. (Stripe-type color tunable OLED. Pixelation

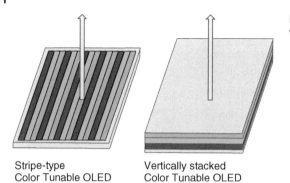

Stripe-type
Color Tunable OLED

Vertically stacked
Color Tunable OLED

Figure 9.11 Two types of color tunable OLED device structures.

can also be used.) The structure requires photopatterning as well as shadow mask patterning to fabricate. On the other hand, the right structure in Figure 9.11 does not need color patterning, but the devices are stacked vertically. (Vertically stacked color tunable OLED. Also called polychromatic OLED.)

Stripe-type needs to mix the color to show arbitrary colors, so diffuser film needs to be attached on the structure. The OLED device can be the same as OLED display, so OLED formulation can be simple.

In the case of vertically stacked color tunable OLED, no diffuser is necessary because the color can be mixed when the light comes out. (So metallic appearance can be made when it is turned off.) It does not require fine patterning; however, it requires very low-resistance transparent conductor technology for intermediate electrode, to avoid the voltage drop of the device.

Figure 9.12 shows the structure of vertically stacked color tunable OLED device [12, 13]. As red, green, and blue OLED devices must be independently controlled, electric current is applied to intermediate electrode. Therefore, anode, intermediate electrode, and cathode electrode must have low enough resistance to avoid the voltage drop. Also to have high transmittance, each electrode needs to have low optical absorption. Eventually, vertically stacked

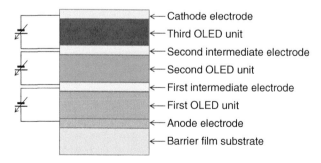

←— Cathode electrode
←— Third OLED unit
←— Second intermediate electrode
←— Second OLED unit
←— First intermediate electrode
←— First OLED unit
←— Anode electrode
←— Barrier film substrate

Figure 9.12 Structure of vertically stacked color tunable OLED.

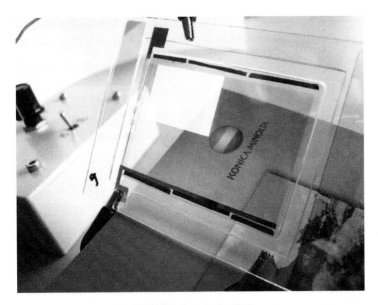

Figure 9.13 Transparent OLED lighting device [12].

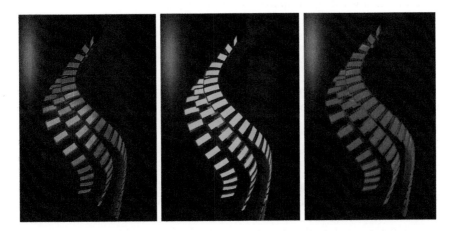

Figure 9.14 Lighting instrument using vertically stacked color tunable OLED lighting [12].

color tunable OLED is a three-unit stack of transparent OLED device. With the same technology, transparent OLED device can be fabricated as shown in Figure 9.13. By using high-transmittance and low resistance transparent conductor, three transparent OLED devices can be stacked successfully to show 16 million color reproduction of vertically stacked OLED devices as shown in Figure 9.14.

The technology can be applied to OLED displays, which may significantly enhance the device lifetime (discussed in Section 5.1.1.3).

9.6 OLED LIGHTING DESIGN

The pixel size in an active-matrix display ranges within hundreds of micrometers, so any voltage drop occurring in a pixel is trivial. However, the dimensions of devices designed for lighting applications are much larger (e.g., within the centimeter range), so voltage drop in these devices can be significant, which may cause a serious luminance uniformity issue.

9.6.1 Resistance Reduction

The light emission side of an OLED electrode (anode electrode in the case of bottom emission structure) must be transparent, so ITO or IZO is used. As these compounds have high resistance values in the range of tens or hundreds of ohms per square, electric current should be supplied from multiple feeding points to ensure better uniformity. Moreover, the feeding points should be connected to the OLED panel pad via low-resistance (e.g., Al or Ca) wiring. (See Section 6.4.3 for voltage drop for AMOLED displays.)

Also, regarding the transparent electrode, the process should be optimized to ensure that resistivity is as low as possible and that there is good transparency.

9.6.2 Current Reduction

To reduce the electric current, several approaches are generally used in combination, including the following:

1. Current efficiency enhancement
2. Tandem device design
3. Serial connection of an OLED device.

Current efficiency enhancement can be achieved by material change, such as improving the emission capability using phosphorescent material [1, 2] (discussed in Section 2.4.2.3). Also, light extraction enhancement, discussed in Section 9.4, can reduce the electric current necessary to obtain the same panel luminance and also provide better luminance uniformity (thus eliminating mura) by reducing the lower voltage drop due to low driving current.

Use of a tandem device structure (discussed in Section 6.4.3) is also an effective approach to decrease the electric current. Theoretically, a two-stack tandem structure reduces the electric current by half, and a three-stack layout reduces the current to a third of its original value. (For better implementation, the light path length should be carefully designed for each emission layer.) Tandem structure also has the merit of extending the lifetime of an OLED device.

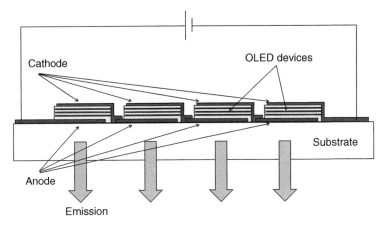

Figure 9.15 Schematic representation of a serial connection structure.

As discussed in Section 6.4.3, N times stack tandem structure results in N^n times the lifetime (two-stack tandem OLED gives 2.82 times and three-stack gives 5.20 times the lifetime, as theoretically discussed). However, creating a tandem structure entails an increased number of process steps, thus increasing the overhead manufacturing cost.

A serial connection structure such as that shown in Figure 9.15 is also used to reduce the electric current. The emission area is partitioned into several areas. By means of the normal OLED fabrication process, OLED devices are fabricated in each partition simultaneously. The cathode electrode area is extended so that it is connected to the anode electrode of the next partition. As a result, the partitions are connected in series, and, therefore, N times light flux is generated with the same current compared with a normal OLED structure. The scenario sounds similar to a tandem structure; however, unlike the tandem structure, the serial structure does not improve lifetime or device longevity because there is also a reduction in the aperture ratio, which offsets the current reduction advantage. It is important to combine all the technologies discussed above in order to design an OLED lighting panel with good uniformity.

9.7 ROLL-TO-ROLL OLED LIGHTING MANUFACTURING

Tsujimura et al. presented large-scale roll-to-roll manufacturing of OLED lighting device in 2014 [14, 15]. It was claimed that flexible OLED has advantages over LEDs in terms of lightweight, thinness, and flexibility. A flexible OLED is capable of forming arbitrary curvature, so could be used for new applications, which conventional point, linear and circular light sources could not make possible (Figure 9.16).

Figure 9.16 OLED Tulip illumination demonstrated in Japanese theme park with shining flexible-OLED petals.

To fabricate a flexible OLED lighting panel, there are two approaches possible:

1. Laser liftoff method (Section 8.1.3)
2. Roll-to-roll method.

Laser liftoff is a method used for OLED displays. The flexible pattern is fabricated on polyimide coated onto the glass, so the pattern size can be very small due to the low thermal expansion of the glass substrate. Accurate patterning can be achieved, so high-resolution display manufacturing is possible. Roll-to-roll manufacturing cannot thus far be applied to displays, leaving the laser liftoff method as the sole option for flexible high-resolution display manufacturing.

On the other hand, OLED lighting does not require high-resolution patterning. Therefore, it is possible to adopt a manufacturing method with higher manufacturability. Roll-to-roll manufacturing has been used in many industries to achieve low manufacturing cost. Of course, the laser liftoff method can still be adopted for OLED lighting manufacturing. However, it was claimed that there are several advantages in roll-to-roll manufacturing compared with the laser liftoff method because laser lift off method has the following features:

1. Laser liftoff method requires very expensive equipment, which adds depreciation cost.

2. The device is released after fabricating all the processing. If the device yield is lost in the release step, this impacts the device costs (variable costs).

3. The polyimide and mother glass are expensive materials, and the glass brings further costs associated with issues in recycling.

4. In the case of roll-to-roll manufacturing, no high cost equipment is necessary for making flexible device, there is no release step to affect yields and no additional cost related to flexible substrates.

References

1 "Solid-State Lighting R&D Plan", U.S. Department of Energy, DOE/EE-1228, (May 2015).

2 M. S. Weaver, P. A. Levermore, V. Adamovich, K. Rajan, C. Lin, G. Kottas, S. Xia, R. Ma, R. C. Kwong, M. Hack, and J. Brown, White phosphorescent OLED lighting: a green technology, *SID 2010 Digest*, p. 208 (2010).

3 T. Tsujimura, K. Furukawa, H. Ii, H. Kashiwagi, M. Miyoshi, S. Mano, H. Araki, and A. Ezaki, World's first all phosphorescent OLED product for lighting application, *IDW 2011 Digest*, p. 455 (2011).

4 T. Komoda, H. Tsuji, T. Nishimori, N. Ide, T. Iwakuma, and M. Yamamoto, High-performance and high-CRI OLEDs for lighting and their fabrication processes, *IDW 2009*, p. 1029 (2009).

5 C. F. Madigan, M.-H. Lu, and J. C. Sturm, Improvement of output coupling efficiency of organic light-emitting diodes by backside substrate modification, *Appl. Phys. Lett.* **76**:1650 (2000).

6 Y.-S. Tyan, Y. Rao, J.-S. Wang, R. Kesel, T. R. Cushman, and W. J. Begley, Fluorescent white OLED devices with improved light extraction, *SID 2008 Digest*, pp. 933–936 (2008).

7 R. Meerheim, M. Furno, S. Hofmann, B. Lüssem, and K. Leo, Quantification of energy loss mechanisms in organic light-emitting diodes, *Appl. Phys. Lett.* **97**(253305) (2010).

8 K. Kato, T. Iwasaki, and T. Tsujimura, Over 130 lm/W all-phosphorescent white OLEDs for next-generation lighting, *J. Photopolym. Sci. Technol.* **28**(3):335–340 (2015).

9 J. Kim, P. Ho, N. Greenham, and R. Friend, Electroluminescence emission pattern of organic light-emitting diodes: implications for device efficiency calculations, *J. Appl. Phys.* **88**:1073–1081 (2000).

10 W. Koo, S. Jeong, F. Araoka, K. Ishikawa, S. Nishimura, T. Toyooka, and H. Takezoe, Light extraction from organic light-emitting diodes enhanced by spontaneously formed buckles, *Nature Photon.* **4**:222–226 (2010).

11 S. Murano, D. Pavicic, M. Furno, C. Rothe, T. Canzler, A. Haldi, F. Loeser, O. Fadhel, F. Cardinali, and O. Langguth, Outcoupling enhancement mechanism investigation on highly efficient PIN OLEDs using crystallizing evaporation processed organic outcoupling layers, *SID Symposium Digest of Technical Papers*, Vol. **43**, Issue 1, pp. 687–690 (2012).

12 T. Tsujimura, T. Hakii, and S. Noda, A color-tunable polychromatic organic-light-emitting-diode device with low resistive intermediate electrode for roll-to-roll manufacturing, *IEEE Trans. Electron Dev.* **63**(1):402–407 (2016).

13 T. Tsujimura, T. Hakii, T. Nakayama, H. Ishidai, T. Kinoshita, S. Furukawa, and K. Yoshida, Development of a vertically-stacked color-tunable polychromatic organic-light-emitting-diode device for roll-to-roll manufacturing, *J. Soc. Inf. Disp.* **24**(4):262–269 (2016).

14 T. Tsujimura, J. Fukawa, K. Endoh, Y. Suzuki, K. Hirabayashi, and T. Mori, Flexible OLED using plastic barrier film and its roll-to-roll manufacturing, *SID 2014 Digest*, Vol. **45**, Issue 1, pp. 104–107 (2014).

15 T. Tsujimura, J. Fukawa, K. Endoh, Y. Suzuki, K. Hirabayashi, and T. Mori, Development of flexible organic light-emitting diode on barrier film and roll-to-roll manufacturing, *J. SID* **22**(8):412–418 (2014).

Appendix

The wavelength values of the color matching function, which is necessary for calculation of the tristimulus value x, y, z from the emission spectrum, are listed in Table A.1.

Table A.1 Color Matching Function λ Values

Wavelength (nm)	$\bar{x}(l)$	$\bar{y}(l)$	$\bar{z}(l)$
360	0.00013	3.92E−06	0.000606
365	0.000232	6.97E−06	0.001086
370	0.000415	1.24E−05	0.001946
375	0.000742	2.2E−05	0.003486
380	0.001368	0.000039	0.00645
385	0.002236	0.000064	0.01055
390	0.004243	0.00012	0.02005
395	0.00765	0.000217	0.03621
400	0.01431	0.000396	0.06785
405	0.02319	0.00064	0.1102
410	0.04351	0.00121	0.2074
415	0.07763	0.00218	0.3713
420	0.13438	0.004	0.6456
425	0.21477	0.0073	1.03905
430	0.2839	0.0116	1.3856
435	0.3285	0.01684	1.62296
440	0.34828	0.023	1.74706
445	0.34806	0.0298	1.7826
450	0.3362	0.038	1.77211

(*Continued*)

OLED Display Fundamentals and Applications, Second Edition. Takatoshi Tsujimura.
© 2017 John Wiley & Sons, Inc. Published 2017 by John Wiley & Sons, Inc.

Table A.1 (Continued)

Wavelength (nm)	$\bar{x}(l)$	$\bar{y}(l)$	$\bar{z}(l)$
455	0.3187	0.048	1.7441
460	0.2908	0.06	1.6692
465	0.2511	0.0739	1.5281
470	0.19536	0.09098	1.28764
475	0.1421	0.1126	1.0419
480	0.09564	0.13902	0.81295
485	0.05795	0.1693	0.6162
490	0.03201	0.20802	0.46518
495	0.0147	0.2586	0.3533
500	0.0049	0.323	0.272
505	0.0024	0.4073	0.2123
510	0.0093	0.503	0.1582
515	0.0291	0.6082	0.1117
520	0.06327	0.71	0.07825
525	0.1096	0.7932	0.05725
530	0.1655	0.862	0.04216
535	0.22575	0.91485	0.02984
540	0.2904	0.954	0.0203
545	0.3597	0.9803	0.0134
550	0.43345	0.99495	0.00875
555	0.51205	1	0.00575
560	0.5945	0.995	0.0039
565	0.6784	0.9786	0.00275
570	0.7621	0.952	0.0021
575	0.8425	0.9154	0.0018
580	0.9163	0.87	0.00165
585	0.9786	0.8163	0.0014
590	1.0263	0.757	0.0011
595	1.0567	0.6949	0.001
600	1.0622	0.631	0.0008
605	1.0456	0.5668	0.0006
610	1.0026	0.503	0.00034
615	0.9384	0.4412	0.00024

Table A.1 (Continued)

Wavelength (nm)	$\bar{x}(l)$	$\bar{y}(l)$	$\bar{z}(l)$
620	0.85445	0.381	0.00019
625	0.7514	0.321	0.0001
630	0.6424	0.265	5E−05
635	0.5419	0.217	0.00003
640	0.4479	0.175	0.00002
645	0.3608	0.1382	0.00001
650	0.2835	0.107	0
655	0.2187	0.0816	0
660	0.1649	0.061	0
665	0.1212	0.04458	0
670	0.0874	0.032	0
675	0.0636	0.0232	0
680	0.04677	0.017	0
685	0.0329	0.01192	0
690	0.0227	0.00821	0
695	0.01584	0.005723	0
700	0.011359	0.004102	0
705	0.008111	0.002929	0
710	0.00579	0.002091	0
715	0.004109	0.001484	0
720	0.002899	0.001047	0
725	0.002049	0.00074	0
730	0.00144	0.00052	0
735	0.001	0.000361	0
740	0.00069	0.000249	0
745	0.000476	0.000172	0
750	0.000332	0.00012	0
755	0.000235	8.48E−05	0
760	0.000166	0.00006	0
765	0.000117	4.24E−05	0
770	8.31E−05	0.00003	0
775	5.87E−05	2.12E−05	0
780	4.15E−05	1.5E−05	0
785	2.94E−05	1.06E−05	0

(*Continued*)

Table A.1 (Continued)

Wavelength (nm)	$\bar{x}(l)$	$\bar{y}(l)$	$\bar{z}(l)$
790	2.07E−05	7.47E−06	0
795	1.46E−05	5.26E−06	0
800	1.03E−05	3.7E−06	0
805	7.22E−06	2.61E−06	0
810	5.09E−06	1.84E−06	0
815	3.58E−06	1.29E−06	0
820	2.52E−06	9.11E−07	0
825	1.78E−06	6.42E−07	0
830	1.25E−06	4.52E−07	0

Index

a

acceleration factor   48, 50
active-matrix OLED display   125
Adobe-RGB standard   113
AFM   173
Al   35
AlNd   172
Al_2O_3 240
Alq_3 14, 35, 44, 69
alternating magnetic field
   crystallization (AMFC)   205,
   206
ambient light reflection   100
AMOLED   8, 125, 203, 217, 272
amorphous silicon (a-Si)   167, 172,
   185, 203, 205, 207, 232, 236
anode line structure   132, 205
anthracene   7, 12
anti-bonding orbital   15
aperture adjustment   48
aperture ratio   48, 177, 185, 192, 222,
   273
area light source   102, 104
area source   72
atomic layer deposition (ALD)   86, 87,
   209

b

backlight unit   99
backplane substrate   125
BAlq   46

bandgap

bandgap   63
band heater   67
bank   152, 155
barrier   88, 240
barrier lowering   26
bathocuproine (BCP)   65
biradical   40
blackbody locus   117
blackbody radiation   117, 259
black dot   79
black spot   79
Bloch wave   45
BN   73
boat   67, 68, 72
bonding orbital   21
boron nitride   73
bottom-contact bottom-gate TFT
   244
bottom emission   45, 81, 185, 189,
   192, 245, 272
bottom-gate TFT   169, 244
Bragg reflector   45
BT-2020   113

c

calcium test   240
calculus of variations   17
CaO   88
cap glass   82, 153
cap metal   82
carrier balance   36, 67

OLED Display Fundamentals and Applications, Second Edition. Takatoshi Tsujimura.
© 2017 John Wiley & Sons, Inc. Published 2017 by John Wiley & Sons, Inc.

cathode separator   121
charge generation layer (CGL)   54
charge injection   7, 24, 32
charge transfer complex   53
chemical vapor deposition (CVD)
      153, 160, 169, 172, 204, 240
chroma   116
CIE1931 color space   111
CIE1964 color space   257, 259
CIE1976 color space   114
CIE-LAB uniform color space   115
CIE-UV uniform color space   115
circular polarizer   100, 125, 137
CMOS   169, 171, 199
cohesion coefficient   69
cold-cathode fluorescent tube, (CCFL)
      99
color boosting   119
color conversion medium (CCM)   149
color difference   47, 116, 258, 259
color filter   136, 148, 174, 189
color gamut   106, 119
color matching function   109, 277
color rendering index (CRI)   259, 263
color reproduction   101, 109, 119,
      255, 259
color tunable OLED   269
common-anode   131
common-cathode   131
compensation circuit   190, 205
concentration quenching   38, 63, 108
conduction band   32
conduction band minimum (CBM)
      207
connector layer   54
contrast ratio   100, 105, 136, 215
copper phthalocyanine (CuPc)   35,
      242
correlated color temperature (CCT)
      260
Coulomb integral   18
coverage   86
critical angle   43

crosstalk   124, 183
CRT   8, 111
crucible   68, 73, 91, 158
crystal thickness monitor   79, 161
CT complex   53
current efficiency   106, 107, 184, 189
current programming   192
current–voltage curve   95
cyclic voltammetry (CV) method   93

d
D65   117
D93   117
dark spot   79
data driver   121, 134, 197
DCI-P3   113
DCJTB   38
delayed fluorescence   37, 40
Dexter energy transfer   39
differential aging   46
diffraction structure   45, 268
diffusion structure   266
diffusion surface   44
digital driving   194
diode laser thermal annealing (DLTA)
      204
DOD   152
doping   38, 47, 63, 73, 160, 168, 170,
      183
driver IC   101, 127, 154, 189, 192, 193

e
EB   68
EBU standard   111, 119
edge cover   121
edge growth   79
edge insulator   121
electric double layer   28
electron affinity   16
electron beam evaporation   75
electron injection layer (EIL)   28, 34,
      63, 64

electron mobility   95
electron pair   23
electron transport layer (ETL)   24, 35, 63, 65
electroplating   146
emission layer (EML)   35, 45, 62, 67, 272
encapsulation glass   154
ENERGY STAR program   259
etching   144, 176, 205
etch selectivity   177
evanescent field   269
evaporation   12, 68, 73, 75, 101, 150, 159
exchange integral   18
excimer   53
excimer laser annealing   170, 179, 190, 200, 207, 248
exciplex   53
exciton   7, 12, 33, 65, 93
exciton decay time   24
exciton generation efficiency   36
EXIF 2.2   113
external compensation   193
external extraction structure (EES)   267
external quantum efficiency (EQE)   36, 267

f

face sealing   87
facing-target   245
Fermi–Dirac distribution function   25
Fermi–Dirac statistics   23
flash evaporation   158
Flash Mask Transfer Lithography (FMTL)   151
flexible display   235, 241
fluorescence emission   8, 23, 39
fluorescence quantum efficiency   36
FMTL   151
Förster energy transfer   39
Fowler–Nordheim current   28

Franck–Condon principle   14
FTS   245
full-color display   134

g

gas frontal chromatography (GFC) column   94
gel permeation chromatography (GPC) column   94
glass substrate   236
glass transition temperature   240
G-line   176
global mura compensation (GMC)   193, 219, 232
GPU   113
gradual-channel approximation   182, 245
graphic engine   113
graphite   73
ground state   17

h

heavy-metal effect   24, 38, 63
Helmholtz–Kohlrausch effect   106
hole blocking layer (HBL)   35, 51, 65
hole injection layer (HIL)   35, 61
hole transport layer (HTL)   24, 35, 62
HOMO   15, 24, 29, 34, 35, 61, 93, 173, 244
hot-wall method   158, 163
HPLC   93
HTO method   90
hue   115
humidity   79, 84, 89
hyperacuity   247

i

IGZO   207, 209
illuminance   101, 103, 104, 137
image burning   46, 114
image force   26
image potential   27
initial decay   47

injector   94, 160
inkjet method   155
inorganic EL   11
integrated circuit   125, 197
intermediate layer   54
internal conversion   15, 37
internal extraction structure (IES)
        266
internal quantum efficiency (IQE)   36
intersystem crossing   37, 51
invar metal   144
inverse photoemission spectroscopy
        (IPES)   93
inverted staggered TFT   167
ionization energy   91
ionization potential   16, 35, 91
ITO   35, 81, 173, 245, 272
I–V curve   95
IZO   172, 272

j

Jablonski energy diagram   21, 23
JPEG   111

k

Knudsen cell (K cell)   69

l

Labyrinth effect   84
Lambertian distribution   106
laser-induced patternwise sublimation
        (LIPS)   151
laser lift off   236, 274
layer-by-layer method   86, 204
LCD   8, 99, 125, 136, 153, 187, 190,
        229, 248
leakage current   170, 183, 242
LED   8, 11, 99
LiF   28, 35, 63
lifetime   12, 24, 34, 46, 48, 67, 114,
        116, 124, 147, 159, 189, 192,
        240, 262, 272
light extraction   43, 262, 268, 272

light guide   99
lightly doped drain (LDD)   170, 184
lightness   115
light-to-heat conversion (LTHC) layer
        149
linear region   180, 183, 194
linear source   75
LITI method   149
L-I-V curve   108
living room contrast ratio   105, 136
location-controlled polysilicon   205
low-temperature polysilicon (LTPS)
        154, 178, 183, 190, 193, 197,
        200, 207, 236
luminance   8, 36, 39, 46, 99, 100, 101
luminance decay   46
luminance–voltage curve   108
luminous efficacy   107
luminous exitance   103
luminous flux   101, 102, 103, 104
luminous intensity   44, 101, 104
LUMO   16, 24, 29, 61, 64, 91

m

manifold   160
maximum luminosity factor   107
Maxwell–Boltzmann distribution   26
McAdam ellipse   114, 230, 260
mean free path   69
melting-type material   68, 73
metal foil substrate   240
metal halide   28
metal induced lateral crystallization
        (MILC)   178, 205
MgAg   35, 245
MIC   178, 205
microcavity effect   45
microcrystalline silicon   203
microlens array   45, 200, 266
micropyramidal structure   45
minimum angle of resolution (MAR)
        247
MISFET   181

mobility   29, 33, 167, 180, 183, 192, 202, 207, 244, 245
Mocon method   89
molecular orbital   15
MOSFET   180, 245
MoTa   172
MoW   172
multiline scanning   124
multi-photon   54
multiple trapping and release (MTR) model   33
Munsell color system   116
mura   190, 272

n

N^+ amorphous silicon   172
N-channel   169, 184
negative-type resist   173
NIBS   209
NMOS   169, 199
nozzle coating   155
NPB   35
NQD   174
NTSC%   111

o

organic TFT   241, 245
oscillation frequency   77
outcoupling efficiency   36
OVPD   158, 164
oxide semiconductor   169, 181, 207, 232
oxide TFT   242
oxygen   70, 84, 88, 89, 94, 173, 240

p

Partial Laser Anneal Silicon (PLAS)   202
particle   23, 148, 174
passive-matrix OLED (PMOLED)   8, 121
Pauli's exclusion principle   21
P-channel   169, 191

PECVD   169, 203
PEN   240
PenTile   138
PEP   173
perfect diffusion   105
perimeter seal   81
permeability   240
perturbation   17
perylene   47
PET   240
phase shift   44
phosphorescence quantum efficiency   36
phosphorescent dopant   63
phosphorescent emission   8, 23, 36, 52
phosphorescent material   63, 272
photoelectron   91
photoemission spectroscopy (PES)   92
photoluminescence (PL)   10, 12
photon   33, 54
photonic array   45
photopatternable PLED   153
piezoelectric device   155
π bonding   15
π^* bonding   15
$\pi-\pi$ interaction   29
Pirani gauge   79, 161
pixel definition layer (PDL)   121
pixel shrinkage   79
planarization layer   176, 236
Planckian locus   117, 260
Planck radiation law   117
PLED   8, 13, 153, 157
PMOS   154, 169, 199
Pointer's color   113
point light source   44, 102
point source   72, 159
Poisson distribution   217
Poisson equation   31
polarizer   99, 125, 136
polaron   40

polychromatic OLED   270
polycrystalline silicon   167, 205
polymer substrate   235
polysilicon TFT   169, 172
polyvinylcarbazole (PVK)   14
Poole–Frenkel conduction   32, 245
positive-type resist   173
potential barrier   25, 61
power efficacy   55, 106, 107, 268
purification   67
PVD   172

q

quartz   73, 77
quartz oscillator   77
quartz tube   67
quencher   40, 108

r

Rec709 standard   111
relative luminosity curve   107, 229
resistive heating method   72
revolver   73
RGB side-by-sidemethod   148
RGBW   138, 221
Richardson constant   26
Richardson–Dushman equation   26
Richardson effect   25
RIST method   151
rolloff phenomenon   108
roll-to-roll   236, 273
rubrene   38

s

S_1 21
saturation region   183, 194, 245
scalable technology   217
scan driver   199
Schottky effect   26
Schottky plot   27
Schottky thermionic emission   24, 25, 32
SCLC   29

self-aligned doping   168
serial connection structure   273
shadowing effect   143
shadow mask patterning   143
Short Thermal Exposure source   165
silane   172
silicon nitride (SiN_x)   172, 268
silicon oxide   172
Si_3N_4 240
singlet exciton   40
singlet state   20
SiO_xN_y 240
SLS method   200, 232
SMOLED   8
SMPTE-170M   117
Snell's law   43
solid angle   70, 105
solid-phase crystallization (SPC)   177, 205
source-follower   132
space-charge-limited current   29
spatial ALD   86
SPC   177, 219
spin function   20
spin-orbital interaction   38
spin–orbit coupling   63
SPP wave   268
sputtering   172, 209, 245
s-RGB   113
staggered TFT   167
Stern–Volmer plot   42
sublimation-type material   68, 73
subpixel   47, 109, 116, 121, 138, 229
subpixel rendering   138
substrate mode   42, 43, 262, 267
superamorphous silicon   202
surface energy   156
surface plasmon   43, 269
surface-plasmon polariton wave (SPP wave)   268
surface reflection   44, 105, 136
s-Ycc standard   113

t

T_1 21
T_{50} 47
T_{70} 47
T_{95} 47
T_{97} 47
TADF-assisted fluorescence (TAF)
        53
tandem OLED   55, 189, 273
target–source distance   72
2T1C circuit   129, 190
thermal decomposition   159
Thermally Activated Delayed
        Fluorescence   51
thermoball   75
thickness measurement   76
thin-film transistor (TFT)   68, 76, 101,
        108, 125, 126, 153, 154, 167,
        169, 180, 183, 189, 190, 194,
        197, 200, 204, 205, 236, 244,
        245
threshold voltage   183, 185, 191, 209,
        245
TMAH   174
top-contact bottom-gate TFT   243
top emission   45, 82, 83, 192, 245
top-gate TFT   169
Tortuous path effect   84
transparent cathode   245, 247
trapping state   32, 33
triplet exciton   40
triplet state   20
triplet–triplet annihilation (TTA)   40,
        52
triplet–triplet fusion (TTF)   40
tristimulus value   111, 117, 258, 277
T/S   72
tungsten filament   73

tunneling injection   24, 28
turret   73

u

UV   12, 83, 91, 115, 153, 174, 202
UV-O_3 treatment   173

v

vacuum-level shift   24, 28, 29
vertically-stacked color tunable OLED
        270
vertical transition   14
vibrational levels   15
viewing angle   121
viewing condition   121
viewing distance   247
VIST   158
visual acuity   247
voltage drop   185, 189, 272
voltage programming   190, 203

w

wavefunction   20
waveguide mode   42, 267
white + color filter method   148, 149,
        219, 224, 230
white point   48, 116, 117, 226, 260
workfunction   7, 12, 27, 35, 61, 64, 91,
        173
W-RGBW   230
WVTR   88, 240

y

Yield simulation   217

z

ZnO   207
Z ratio   78

Wiley-SID Series in Display Technology

Series Editors:
Anthony C. Lowe and Ian Sage

Physics and Technology of Crystalline Oxide Semiconductor CAAC-IGZO: Application to Displays
By Shunpei Yamazaki, Tetsuo Tsutsui

Physics and Technology of Crystalline Oxide Semiconductor CAAC-IGZO: Application to LSI
By Shunpei Yamazaki, Masahiro Fujita

Physics and Technology of Crystalline Oxide Semiconductor CAAC-IGZO: Fundamentals
By Shunpei Yamazaki, Noboru Kimizuka

Modeling and Optimization of LCD Optical Performance
By Dmitry A. Yakovlev, Vladimir G. Chigrinov, Hoi-Sing Kwok

Fundamentals of Liquid Crystal Devices, 2nd Edition
By Deng-Ke Yang, Shin-Tson Wu

Addressing Techniques of Liquid Crystal Displays
By Temkar N. Ruckmongathan

Interactive Displays: Natural Human-Interface Technologies
By Achintya K. Bhowmik

Illumination, Color and Imaging: Evaluation and Optimization of Visual Displays
By P. Bodrogi, T. Q. Khan

OLED Display Fundamentals and Applications
By Takatoshi Tsujimura

3D Displays
By Ernst Lueder

Liquid Crystal Displays: Fundamental Physics and Technology
By Robert H. Chen

Transflective Liquid Crystal Displays
By Zhibing Ge, Shin-Tson Wu

Liquid Crystal Displays: Addressing Schemes and Electro-Optical Effects, 2nd Edition
By Ernst Lueder

LCD Backlights
By Shunsuke Kobayashi (Editor), Shigeo Mikoshiba (Co-Editor), Sungkyoo Lim (Co-Editor)

Introduction to Flat Panel Displays
By Jiun-Haw Lee, David N. Liu, Shin-Tson Wu

Projection Displays, 2nd Edition
By Matthew S. Brennesholtz, Edward H. Stupp

Photoalignment of Liquid Crystalline Materials: Physics and Applications
By Vladimir G. Chigrinov, Vladimir M. Kozenkov, Hoi-Sing Kwok

Mobile Displays: Technology and Applications
By Achintya K. Bhowmik (Editor), Zili Li (Co-Editor), Philip J. Bos (Co-Editor)

Introduction to Microdisplays
By David Armitage, Ian Underwood, Shin-Tson Wu

Fundamentals of Liquid Crystal Devices
By Shin-Tson Wu, Deng-Ke Yang

Polarization Engineering for LCD Projection
By Michael D. Robinson, Gary Sharp, Jianmin Chen

Flexible Flat Panel Displays
By Gregory Crawford (Editor)

Digital Image Display: Algorithms and Implementation
By Gheorghe Berbecel

Display Interfaces: Fundamentals and Standards
By Robert L. Myers

Colour Engineering: Achieving Device Independent Colour
By Phil Green (Editor), Lindsay MacDonald (Editor)

Reflective Liquid Crystal Displays
By Shin-Tson Wu, Deng-Ke Yang

Display Systems: Design and Applications
By Lindsay MacDonald (Editor), Anthony C. Lowe (Editor)

Printed and bound by CPI Group (UK) Ltd, Croydon, CR0 4YY

27/10/2024

14580671-0001